HOW THE ZEBRA GOT ITS STRIPES

HOW THE ZEBRA GOT ITS STRIPES

Darwinian Stories Told
Through Evolutionary Biology

LÉO GRASSET

PEGASUS BOOKS
NEW YORK LONDON

How the Zebra Got Its Stripes

Pegasus Books Ltd
148 West 37th Street, 13th Floor
New York, NY 10018

First Pegasus Books hardcover edition May 2017

ISBN: 978-1-68177-414-5

10 9 8 7 6 5 4 3 2 1

Printed in the United States of America
Distributed by W. W. Norton & Company, Inc.

Contents

PART I

Evolution in its Guises

'Nothing in biology makes sense except in the light of evolution,' wrote the eminent geneticist Theodosius Dobzhansky. But sometimes, when the questions being explored are at the cutting edge of scientific research, the light of evolution casts shadows that are hard to decipher!

Chapter 1

The Female Hyena's Penis

WHY DO COWS HAVE HORNS? Why is it that most small antelope females do not have them? Why do men have nipples? Why does a female hyena have a clitoris that is, to the naked eye, indistinguishable from a male hyena's penis?

More generally, the question is this: why do some morphological characteristics that appear to have a function exclusive to one sex also exist in the other? Nipples are a good example: in women they serve to suckle infants, grouping the milk ducts together and providing an interface between the baby's mouth and the mother's mammary glands. But in men their function isn't clear: what is the point of a pair of nipples if they're not for feeding a baby? Perhaps we should simply say, 'Why not? Does everything have to have a function?'

Of the laws that govern the evolution of living beings, selection is the most powerful. This, as we'll see, is especially evident on the great African savannah. If one individual possesses a slight advantage over another, it will produce more young. If these offspring inherit and pass on the same advantage, the descendants of the advantaged individual will eventually dominate the species' gene pool, while those of its erstwhile rival will be consigned to evolutionary oblivion. This, of course, is a simplification: in the real world things are never that straightforward. But to illustrate the theory, let's turn back to the evolution of the nipple.

Suppose that, in the beginning, all men have flat pectorals. Then, one day, a man appears sporting a pair of nipples which emit an intoxicating pheromone. This scent has such a seductive effect on the women he encounters that he fathers 50 per cent more children than his nippleless rivals. If his aphrodisiac nipples are heritable, the children of this fortunate mutant will also be able to sire 50 per cent more offspring, who in turn will go on to produce 50 per cent more great-grandchildren for the mutant and so on down the generations. After five centuries, or twenty generations of twenty-five years, the 'nipples and pheromones' human will have some 3,325 times (1.5^{20}) more descendants than the flat-pectorals type: a colossal difference. As long as it can be inherited, even the slightest advantage in the number of offspring will have major repercussions down the generations. Minor effects become cumulative, and in this case, would eventually result in a human population in which all males were equipped with nipples.

In this view of the world, if an organ exists it must have a function. If it appears to be redundant, it is only because we have not yet discovered what that function is. Biologists with this turn of mind might propose that women prefer men who have nipples to men who do not. Or they might suggest a social function: in mothers we know that as a baby suckles it releases a surge of the hormone oxytocin, which promotes feelings of wellbeing and social cohesion and is thought to strengthen the bonding process between mother and baby. In other words, more suckling equals more love for infants, who therefore have better survival rates, which means more babies. These explanations and others like them derive from the belief that everything exists for a reason.

But another school of thought proposes that there are possible evolutionary scenarios in which the male nipple has no practical function at all. All human embryos start out as female: the female sex is the basic form from which the male sex will differentiate itself. The first male hormones do not appear until the eighth week of pregnancy. In other words, the male embryo has to make its male organs using the material available to it, which already tends to the feminine. As the nipples are present from the sixth week, the male embryo is stuck with them. At this point you can reverse the logic: every exclusively male characteristic is an additional attribute, hard won by means of major surges of testosterone and bursts of androgens – the classic male hormones. If a female attribute lingers on, and doesn't get in the way or put the male at a disadvantage, it will stay put.

Stringent selection could circumvent this constraint, of course, and drastically favour the male without nipples over his rival who has them but, as this is clearly not the case, we have no reason to lose them.

Understanding the factors resulting in a characteristic that appears extraneous is a challenge for biologists. Here are two more examples, both from the savannah: the penis-shaped

Viewed with the naked eye, the female hyena's clitoris
is indistinguishable from the male's penis.

clitoris of the female hyena, and the horns of the female buffalo.

No, you did not misread that: the female spotted hyena (*Crocuta crocuta*, photograph 2) has a clitoris shaped like a penis. It is known in scientific jargon as a pseudo-penis. An imitation, basically, but a seriously good one. In fact, the female hyena imitates the male genitalia in their entirety: she also has a faux scrotum and faux keratin spines (a very widespread characteristic in mammals) on her faux penis, and not only can she get an erection with her clitoris but she also urinates through it. She has no separate vaginal entrance: her entire external genitalia take the form of a male penis. With the naked eye, it is hard to tell the difference between the male hyena and the female.

When it comes to giving birth, this organ causes the female serious difficulty, because her offspring enter the world via this narrow pseudo-penis. As a result, 15 per cent of mothers die during their first labour, and no fewer than 60 per cent of hyena cubs die at birth. From the evolutionary point of view, therefore, there has to be a pretty persuasive upside to justify the presence of this organ. One advantage is that it is difficult for the males to mate with the females by force; even when she is willing. It will take several attempts before the couple manages to find the right position, because he has to insert his penis into her pseudo-penis. For hyenas, successful mating is a whole art in itself, demanding a degree of expertise from the male and so allowing the female all the time she needs to choose her preferred partner.

For a long time it was thought that the female hyena's pseudo-penis was a consequence of the social hierarchy among hyenas: the females are dominant over the males (they are bigger, which helps), and the more aggressive females dominate the sisterhood. Their aggression is controlled by male hormones, and it used to be thought that the struggle for dominance released a higher level of androgens in the females, leading to the 'accidental' appearance of male organs.

This tortuous explanation – in a nutshell, 'aggressiveness equals androgens equals male organs' – no longer holds water, as we now know that androgens play no part whatever in the appearance of the pseudo-penis. There must be another explanation.

The female genitalia are faithful copies of the male ones, and some theorise that the organ's mimicry is too perfect to be simply a hormonal accident. Instead, they believe that the sexes' resemblance is the result of natural selection, possibly to cut down on rivalry among females. As yet, there is no consensus over the reason for this bizarre mimicry; our only certainty is that there must have been a very strong process of selection underlying it.

Continuing this exploration of bizarre sexual attributes,

consider the behaviour of male impalas (*Aepyceros melampus*) in the rutting season. These antelopes of the bovid family use their horns in aggressive displays, clashing them together violently as they fight for dominance over a harem of as many as a hundred females (photograph 15). The females of this species do not have horns, the sole purpose of which is generally agreed to be for competition among males. Strikingly, the horns point backwards, showing that the goal of these confrontations is not to kill or wound the opposing male, but simply to give him a hefty shove. The function of the horns is thus the same as in numerous other members of the deer family: to provide the males with appendages used primarily for thumping each other and for demonstrating to the females which of them is the most deserving of their attentions. In some deer species horns and antlers can attain dimensions that defy belief. In an example of evolution being pushed to the limits, the extinct Irish elk (*Megaloceros giganteus*) boasted antlers with a span of up to 3.6 metres and weighing as much as forty kilos.

But there are other bovine species in which the females do have horns, such as the African buffalo (*Syncerus caffer*, photograph 5) and the domestic cow. As is often the case, ingenious biologists have come up with several different hypotheses to explain why. One might suspect a 'genetic correlation' between males and females, meaning that males and females are built to the same blueprint (at the start of life, at least), and that this similarity is maintained subsequently. But although such an explanation might hold good for male nipples, it is not the reason that some female ungulates have horns and others do not.

Instead, there is another explanation, for once a fairly simple one. It is that females need horns as a defence when they are unable to hide from predators. Impala coloration serves as camouflage in the tall grass, for instance, whereas buffalo are very large, very dark and very visible. When she is hunted, the female buffalo cannot hide, and her only option is to fight for her life. Unlike those of the impala, the buffalo's horns are necessarily lethal weapons. Not for nothing is the animal known in Africa as the 'widow-maker' or the 'black death': they are responsible for over 200 human fatalities a year across the continent. The moral of this story is that if you are too big to hide, you have to be able to defend yourself.

To summarise: men's nipples seem to have no function; their presence is generally explained as the result of a foetal developmental constraint that is too strong and an evolutionary selection process that is too weak to eliminate them. In the case of the hyena, it seems there must be a very strong selection process for the sexual characteristics of the female to mimic those of the male and that this could be associated with a reduction of conflict within the social group. In large bovines the females have horns, although among other ungulates this characteristic is typically male. The presence of horns in the females has been strongly selected as a defence against predators. In short, apparently 'natural' characteristics, such as male nipples, do not necessarily have a function whereas others that at first glance seem to serve no purpose are in fact the fruit of an important process of selection and very definitely do have a function.

Evolution is a complicated phenomenon: it makes organs and appendages disappear, creates new ones and repurposes existing ones for different functions. In the face of these constant changes, it is sometimes difficult for biologists to understand the functions of the shapes and appearances of the creatures they study: they are all too ready to put forward multiple hypotheses, some of which are contradictory. Perhaps researchers are looking for simple explanations, whereas the exuberant creativity of evolution requires something far more complex.

Chapter 2

The Giraffe's Long Neck

THE CHALLENGE OF EVOLUTIONARY BIOLOGY is to explain the origin and function of adaptations. Sometimes things can prove to be more complicated than they appear at first sight. One example is the giraffe, or rather its improbably long neck. It appears obvious: the giraffe's neck, which can grow to as much as two metres in length, has been selected because it gives its owner exclusive access to the topmost leaves of the trees, and no other animal can reach them. This, then, is an adaptation designed to avoid competition for food with other animals. For many years this was the accepted version – after all, Charles Darwin, no less, touched on the question in *The Origin of Species*. Darwin explained that the species had obtained its very long neck by small, successive stages, each individual with a slightly longer neck being able to survive on average a little better than their shorter-necked relatives:

The giraffe, by its lofty stature, much elongated neck, fore legs, head and tongue, has its whole frame beautifully adapted for browsing on the higher branches of trees. It can thus obtain food beyond the reach of the other ungulata or hoofed animals inhabiting the same country; and this must be a great advantage to it during dearths .... So under nature with the nascent giraffe, the individuals which were the highest browsers and were able during dearths to reach even an inch or two above the others, will often have been preserved.

Subsequently, the giraffe's neck became a textbook case, featuring as an example of natural selection in numerous books and popular articles. In the mid-1990s, however, some biologists raised a major objection to this argument: observations suggested that giraffes did not use their long necks much at all to browse at heights. In fact, at times when competition for food was fiercest, the females could spend up to half their time with their necks held horizontally rather than exploiting their height advantage. These biologists put forward a different scenario, one which revolutionised the classic view of the evolutionary history of the giraffe.

The purpose of a giraffe's neck, they argued, is primarily as a weapon to be wielded in fights between males, just as a male antelope uses its horns or a stag its antlers. Male giraffes indulge in bouts of 'neck fighting' to gain access to the females, swinging their necks at each other violently and using their heavy heads as coshes. The male's skull is extremely thick, and when used as a weapon is capable of breaking vertebrae: the Republic of Niger has only a tiny giraffe population, but in 2009 it recorded

The giraffe's neck serves a number of functions:
which of these functions shaped its evolution is the
subject of much debate among biologists.

two deaths following bouts of neck fighting. In this context, it is clearly an advantage for a male giraffe to have a thicker neck than its rival, and a longer neck provides greater flexibility and torque, thereby making it a more effective weapon. The males that reproduce most successfully also have the longest necks, and so the evolution of the giraffe's neck turns, quite literally, into a tall story.

If this is the case, however, why are female giraffes' necks also long? The only explanation so far offered is that it might be a case of 'genetic correlation between the sexes' – the hypothesis that is so often dragged out when no other explanation seems to fit. Although persuasive, this idea does not explain things very well. If sexual selection is the cause, males should have noticeably longer necks than females, but a study in 2013 found that males' necks were only slightly longer than the females', a difference too small to be explained by sexual selection alone.

To confuse matters further, a study undertaken in 2007 had

concluded that giraffes do indeed use their necks to graze the topmost branches. The researchers fenced off some trees with wire netting so that smaller herbivores could not reach the lower branches, but giraffes could still graze by reaching over the top of the netting. When the fenced-off trees were compared with unfenced trees, the researchers found that the giraffes did, in fact, browse the higher branches when other species were competing for the lower leaves. So perhaps Darwin was right after all: giraffes use their long necks in order to avoid competition. Fossil evidence supplies further backing for his hypothesis: it appears that giraffes developed their long necks between fourteen and twelve million years ago, a period during which Africa underwent a general aridification and its forests gave way to savannah. As the number of trees diminished, competition for each tree must have increased, so favouring the selection of a long neck.

Fortunately, one explanation does not exclude the other: the ability to graze the higher branches is probably an advantage that shaped the evolution of the long neck for both sexes, while its use as a cudgel in competition between males is an evolutionary factor that explains the significant difference in thickness between male and female skulls. In summary, the giraffe's neck has a number of uses, and it can be difficult to say which of them has most strongly influenced its evolution.

In addition, field biologists have proposed a flurry of other hypotheses to explain the neck's elongation. Perhaps the lofty view it affords helps the animals spot predators, or maybe its large surface area assists in regulating body temperature. It has even been suggested that the neck might have evolved in response to giraffes' legs getting longer, so ensuring that they could continue to drink at waterholes.

The evolution of the giraffe's neck shows the range of methods employed by scientists in their attempts to trace the evolutionary history of an adaptation. Over the past 140 years Darwin and his heirs have proposed a variety of rival theories. After painstaking fieldwork and passionate argument, some of these have been judged more favourably than others. For the moment, anyway. The question of the evolution of the giraffe's neck looks set to keep researchers busy for a while yet.

Chapter 3

The Random Flight of the Gazelle

ALL TOO OFTEN you hear commentaries like this on documentaries about the African savannah:

Toki is a patient cheetah. He crouches low among the tall grasses of the African savannah and silently slinks forward towards his prey, a young Thomson's gazelle. Slender and elegant, Toki fixes his eyes on the antelope and can already feel his powerful fangs sinking into its flesh. Suddenly he pounces, his sinewy muscles consuming his energy as he slips through the grasses at the incredible speed of 90 kilometres per hour.

You turn off wearily. You already know about gazelles and cheetahs: the cheetah runs like the wind; the gazelle executes breathtaking leaps as it tries to escape; and the chase is set against the parched grasses of the savannah. But in fact the *way* the Thomson's gazelle evades a cheetah can open up new doors for us, offering us a glimpse of a world that is still largely terra incognita for biologists, and that might prove to be the crucible for a future revolution in science.

Let's look again. The cheetah pounces, stretches and attempts to sink its claws into the unfortunate gazelle. But the gazelle is cunning, and starts running in all directions. Literally. Running in a straight line would guarantee a swift death, as it is no rival for the cheetah when it comes to speed. But if it abruptly changes direction every five seconds it can disorientate the cheetah, disrupt its graceful rhythm, shatter its sprinter's ego into a thousand tiny pieces – or at least gain a chance of surviving. The gazelle's flight is completely random: researchers have found that predicting the next swerve of a gazelle in flight is impossible, and that this is a highly effective strategy in ensuring its survival when attacked. This is not an isolated example of an animal exploiting chance to its advantage: many other animals also resort to this type of behaviour when fleeing attack, and some even use it for finding food and in reproduction. It is technically called 'protean' behaviour, from the Greek sea god Proteus who was able to change his shape rapidly – from water into a lion, a snake, or even a tree.

The existence of random forms of behaviour may seem slightly surprising – most views of animal behaviour suggest that it has been optimised by natural selection, and that individuals adopt compromises that tend to the best possible outcomes, for example what is known as 'optimal foraging theory', where predators attack at the 'right moment', or where herbivores migrate on precise dates, or animals gather in coordinated groups. But the fact that a behaviour may be finely optimised does not mean that it cannot also be random:

The gazelle makes sudden and apparently random changes in its direction of flight in order to disorientate its predator.

if a gazelle with an unpredictable trajectory can survive longer and eventually produce more baby gazelles than a gazelle with a predictable trajectory, then evolution will select the 'chancy' trajectory. Chance may therefore underlie adaptation, and despite appearances there is no contradiction between 'optimisation' and 'randomness'. To imagine that something that behaves in an uncertain way is less effective than its predictable equivalent is an irrational cognitive bias, if a very widespread one among humans.

Alain Pavé, a French scientist who employs mathematics to study biology, has coined the term 'biological roulette' to describe an ability to use chance which has been selected in some animals through evolution. The gazelle has evolved a complex of neurons that creates its zigzag flight route. Gazelles adopt an unpredictable course in order to escape their predators, and theirs is not an isolated case among animals. Many other species explore their environment in search of food in a random fashion, because this is the best strategy to adopt if you are not sure exactly where to look. For example, the larva of the common green lacewing (*Chrysoperla carnea*) randomly wanders over leaves until it comes across the aphids that it feeds upon.

Chance is also a common factor in reproduction: many sea creatures, including sea urchins and mussels, simply release their sperm into the water, where small fluctuations in currents may combine to waft them to their goal. Or they may not. Chance processes in reproduction are even more common in plants, many of which pepper the atmosphere with grains of pollen, their equivalent of sperm. Some of these make contact with the ovules (the equivalent of the egg), which they then fertilise to produce seeds. These seeds are themselves often dispersed into the air and carried off at the mercy of chance fluctuations of the environment: a strong breeze may carry them far away, for example, and deposit them in surroundings that are unfavourable for growth (failure), or in which they will flourish (success). Living creatures therefore use chance to survive, to feed themselves and to reproduce.

But the place of chance in the living world is not limited to animal behaviour and plant dispersal systems: it also plays a major part in the biosphere and in the evolution of organisms, and on many different scales – from an ecosystem covering hundreds of square kilometres to the eye of a fly measuring just a few micrometres across.

The Amazon rainforest contains a huge number of plant species, including up to 16,000 varieties of tree. A single hectare of rainforest may support up to 300 different tree types, all of them consuming the same resources: sunshine, mineral salts and water. One general approach to understanding how species are distributed is that of 'ecological niches'. The idea is that species enter into competition for the resources in their surroundings, and that eventually each of them will come to specialise in one group of resources and to occupy its own niche, from which other species will be excluded by competition. But for some time now this explanation has come up against a stumbling block: tropical rainforests. In tropical rainforests there are few resources for trees, and many different tree species compete to use them in the same way. In 2001 the American ecologist Stephen Hubbell proposed the revolutionary idea that tropical rainforests are not constructed just by competition between species, but also by the workings of chance. He started from the principle that different species of tree do not vary a great deal in their ability to propagate themselves, and that on average all species produce pretty much the same number of seedlings. Minor local differences are due to chance, and the fact that the different species are more or less equal can explain not only the very large number of species present, but also their completely random distribution: some rainforests are so varied that two trees of the same species are hardly ever found in close proximity.

Another example, on a much smaller scale, is found in the common vinegar fly (*Drosophila melanogaster*), often misnamed the fruit fly, which has compound eyes made up of many facets or ommatidia. There are two types of ommatidia, one sensitive to warm colours and the other to cold colours. The combination of these facets endows the fly with colour vision, but how can the ommatidia be mixed up effectively in order to produce an even visual image? A simple and low-energy solution is to rely on chance. Thanks to what statisticians call the law of large numbers, the warm- and cold-sensitive ommatidia stand every chance of being evenly distributed, and this is even more the case if their numbers are very high. Since there are several hundred of them, this is a clever ploy that – thanks to chance – is extremely successful.

Ecology and evolution are themselves intrinsically statistical sciences, founded on probabilities. The major laws of these disciplines require any projections to be based on average changes in the population, and it is impossible to predict with certainty the future of any single individual.

For example, 'the kangaroos that jump the highest are the ones selected', should be expressed as 'the average height of jump in the kangaroo population increases over the generations'. But this does not enable us to make any accurate predictions regarding the offspring of a particular individual with an ability to jump unusually high. We might expect it to have lots of offspring, but – the life of a kangaroo being full of ups and downs – we can only propose hypotheses expressed in the form of probabilities. Suppose, for instance, the high-jumping individual belongs to a very small population of kangaroos, cut off in the depths of the Australian outback. Then suppose that a truck driven by a drunk driver ends up careering out of control through the bush. It ploughs into our little group, knocking them down like ninepins, and our high-jumping kangaroo is squashed flat. The gene that enabled him to jump higher than the others, which was an a priori advantage, vanishes from the population through a simple twist of fate. Chance can play an important role in small populations like that of our kangaroo, a phenomenon known as genetic drift. Selection and genetic drift are often described as 'evolutionary forces', and their relative importance really does depend on the size of the population: a small population will be subject to the effects of chance, while the law of large numbers means that a large population will be less susceptible. But what exactly do we mean by 'chance'?

Of the many definitions of chance, the most generally applied might be something along the lines of 'a force that is assumed to cause events but that cannot be foreseen'. Consider the following sequence of events: a flowerpot falls to the pavement in a Paris street, which causes the trumpet player in a group of buskers to play out of tune; this frightens a cat, which jumps into the lap of a smoker, who drops his cigarette, which lands on the passenger seat of passing lorry loaded with chemical products, which ignites a fire in the cab, which makes the driver lose control of the vehicle, which ploughs into a tanker carrying a reagent, which sets off an explosion that razes Paris to the ground. This chain of events would be unpredictable, and common sense would attribute it to the unfortunate vagaries of chance.

It could be argued that if we could garner information about every component of our world we would then be in a position to predict the future of every one of its parts. In this instance, given sufficient information concerning the instability of the flowerpot and the positions of the trumpet player, the cat, the smoker and the two HGVs, a gifted analyst should have been able to predict the catastrophe. This is not a new idea: the French eighteenth-century mathematician and physicist Pierre-Simon Laplace postulated that,

given for one instant an intelligence that could compre-
hend all the forces by which nature is animated and the
respective situation of the beings who compose it ... it
would be able to embrace in the same formula the movements of the greatest bodies of the universe and those of
the lightest atom: for it nothing would be uncertain and
the future, like the past, would be present to its eyes.

If we accepted this view of the natural world, what we call
chance would be merely the measure of our own ignorance.
But there is also another concept of chance and this comes to
us from quantum physics.

Among the least intuitive of phenomena known to humanity is the principle of quantum superposition. In the quantum
world, a particle may exist in a state that does not correspond
to a unique and well-defined classical value. For example, an
object might be neither red nor blue, but its interaction with
other objects will fix it as a specific colour, whether red or blue.
The physicist Erwin Schrödinger illustrated the weirdness of
quantum superposition with the image of a cat shut in a sealed
box, together with a quantum particle in a state of superposition and a monitor that will automatically release a substance
poisonous to the cat when the particle becomes fixed in one of
its two possible states. As long as the box remains sealed, there
is no disturbance or observation that can fix the particle to a
given value, and it stays in a state of superposition. According
to quantum mechanics, the cat is therefore neither dead nor
alive. As soon as an observer opens the box, the particle makes
a 'choice' between its two potential states and the detection
apparatus either does or does not release the poison. As the

slightest disturbance can tip this system towards a 'choice', in a process known as 'decoherence', it is extremely difficult to maintain a particle in a state of superposition, especially at an ambient temperature.

Yet recent experiments show that nature may have found a number of ways of maintaining electrons in a state of quantum superposition. One example may be found in the plant world, at the very heart of photosynthesis. Photosynthesis is beyond all doubt the most important biochemical process in the living world, introducing solar energy into the biosphere and feeding the entire network of food chains that depend upon it: herbivores, predators, parasites – when it comes down to it, all are dependent on the efficiency of photosynthesis. Light is absorbed by chlorophyll-containing molecular 'antennae' that harness photons and transmit them to biochemical reaction centres where their energy is released. Recent research has suggested that these antennae may be able to keep photons in a state of quantum superposition, which would enable them to simultaneously explore the different routes leading to the reaction centres, to 'select' the shortest one, and to maximise energy efficiency.

In other species, some organs involved in navigation and detection of magnetic fields also seem to make use of particles in quantum superposition. Quantum physics might even play a part in the appearance of mutations in DNA – the process on which the permanent renewal of diversity rests – through 'quantum tunnelling', for example. From ecology to quantum effects, biologists are interested in chance for a variety of reasons: because it can be selected through evolution, because it generates philosophical debates about its origins, and also because it is by definition mysterious and intrinsically surprising. Increasingly, researchers are making it central to their work, and are no longer relegating it to the status of mere 'statistical noise'. Chance enables us to explain and understand. Life is a gamble.

Chapter 4

How the Zebra Got its Stripes

THERE ARE NUMEROUS MAMMALS with contrasting stripes or spots on their coats, and these markings serve a wide variety of functions. Many animals, such as leopards and cheetahs, use them as camouflage. Some, such as polecats, honey badgers and other members of the badger family, use them to deter potential predators, 'Don't annoy me, I'm not in the mood.' Some scientists have suggested that the spots on a giraffe's coat might serve to dissipate heat and lower the animal's temperature, or that they might be a way of enabling individuals to recognise each other.

But however ostentatious such morphological traits may be, it is not always easy to know what the purpose of a species' colouring may be. The plains zebra (*Equus quagga*) offers a magnificent example of this conundrum. This animal is covered from end to end with stripes in an astonishing variety of shapes, widths and even colours, ranging from black in

An attempt at taming a zebra in colonial East
Africa, early twentieth century.

adults to light brown in foals. (While we are on the subject,
it appears that the ancestors of the genus *Equus* had stripes,
and this characteristic occasionally crops up again in domestic
horses. A handful of eccentric nineteenth-century colonists,
mostly British, attempted to domesticate the zebra – without
any great success, although photographs survive of aristocratic
Englishmen attempting jumps on fully saddled-up zebras, and
even of Lord Rothschild's famous barouche drawn through
London by two or three pairs of zebras. The problem with
domesticating zebras is twofold: they are often very aggressive,
and they are also much more slender in build than a horse or
donkey, and therefore inappropriate as either mounts or beasts
of burden.)

But to return to the subject: why stripes? We need to clear
up one thing: zebras' stripes are, in fact, white, on a black background. The zebra embryo starts off completely black, before

bands of white begin to appear, caused by an inhibition in the production of melatonin (the protein responsible for the black coloration). These bands eventually produce the stripes that will be unique to that individual throughout its life. You may be surprised to learn that zebras are not absolutely symmetrical: there are important differences between the patterns on their left side and their right side – the same effect can be seen on many tabby cats. As for the biological function of the stripes, this is more complicated. In 2002, zoologist Graeme D. Ruxton compiled a list of no fewer than eight different theories that have been put forward to explain the origins of these unusual patterns. It is worth scanning quickly down the list, if only to gain an insight into why the question can plunge biologists into the depths of despair:

1. The stripes serve as 'dazzle' camouflage for the group. A group of zebras on the move functions as an optical illusion, disrupting the perceptions of predators, especially in distinguishing where one zebra ends and the next begins when they overlap.
2. Stripes offer camouflage in tall grass.
3. Stripes offer camouflage at night.
4. Stripes around the neck form a recognition zone for mutual grooming by members of the group.
5. Stripes enable individual members of the group to recognise each other.
6. Stripes make it more difficult for predators to home in on the zebra when in pursuit.
7. Stripes deter tsetse flies from landing on the animals.
8. Stripes serve to dissipate heat.

Let's take a closer look at the last four of these, which are the ones most often cited.

Zebras' stripes enable mutual recognition

This idea appears logical, as the patterns are unique to each individual. But at the same time wild horses, which have the same social organisation as zebras, recognise each other perfectly well without recourse to stripes. It also seems fairly unlikely that stripes would serve a function that in closely related species is fulfilled without such an innovation, so the explanation probably lies elsewhere.

Zebras' stripes disrupt predator perception

This entertaining hypothesis in fact contains several others (including reason 1 in the list above). It can be shown, for example, that a striped animal will appear larger than it actually is, which complicates matters for the predator which must estimate accurately when and where to strike. In addition to this, stripes disrupt the predator's perception of the speed and direction of its prey, in exactly the same way as the dazzle camouflage painted on warships in order to disrupt the aim of enemy gunners before the invention of radar. In 2013 serious research recommended that fast-moving military vehicles should be painted with zebra stripes in order to disrupt the aim of enemy personnel armed with rocket launchers.

Simulations have recently shown that the stripes of a moving zebra can produce two optical illusions, both of which reverse the perceived direction of movement. The first of these has a stroboscopic effect, of the kind generally seen in movies when revolving objects such as the wheels of a car are filmed.

In certain natural light conditions too, such objects can give the impression of turning slowly, of standing still, or even of turning in the opposite direction. The second is the barber-pole illusion, in which the diagonal stripes seem to drift upwards as the pole rotates. In an article published in 2014 Martin How and Johannes Zanker suggested the diagonal stripes on zebras' coats produce both these optical illusions and that this disrupts the perceptions of predators who are thus inclined to launch their attacks inaccurately and miss their prey.

Zebras' stripes deter flies

This idea has its origins in the observation that tsetse flies and other flying horrors land less frequently on striped objects than on plain ones, and has been tested on horseflies, the females of which feed on blood.

Horseflies possess an ability to see polarised light that enables them to pinpoint puddles with accuracy: the light reflected from the surface of the water is polarised, as may be seen when they are viewed through polarised sunglasses which cut out horizontal light rays such as reflections from the sea or a wet road surface, so enabling us to see what lies under the waves or view the road without being dazzled. Horseflies use this method to detect the pools where they lay their eggs and find their prey: the herbivores that come there to drink.

In the polarised vision of horseflies, however, zebras' stripes create an effective form of camouflage, as the black-and-white stripes reflect the light in different directions and at different intensities. In an experiment in 2013, biologists observed that horseflies did in fact land less often on models of zebras with stripes than on models without stripes. From this, the authors of the experiment concluded that if stripes fulfilled this function, then they must have been selected – at least in part – for this reason. In 2014 Tim Caro, a professor of wildlife biology, and others at the University of California, Davis, published the results of research that seems to confirm this hypothesis. They noted the natural distribution of numerous subspecies of Equidae – the taxonomic family of horses and related animals – and pointed out that regions where horseflies are naturally present coincide closely with regions where members of the family Equidae are striped. Other indicators back this up, including a correlation between the number of stripes on the abdomen of species in a given region and the presence of tsetse flies in that area. So perhaps zebras evolved their stripes as a response to the pesky flies that feed on them.

Zebras' stripes help to dissipate heat

In 2015, in a Royal Society article entitled 'How the zebra got its stripes: a problem with too many solutions', Brenda Larison at UCLA and an international group of scientists set out to update a 1990s hypothesis: this proposed that the function of zebras' stripes was not to escape from predatory lions or avoid unwelcome attentions from horseflies, but rather to reduce the negative effects of heat. The authors show that, out of all the environmental variables, temperature best explains the geographical disparity in stripe thickness. In hotter regions the stripes are thicker and in cooler regions they tend to disappear. An extreme example can be found in the quagga, a now-extinct subspecies of the zebra, which lived in the coolest region of South Africa and had no stripes at all over most of its body.

The authors don't, however, explain the mechanism by

which stripes confer on their bearer the ability to avoid heat. They propose two hypotheses. First, it may be that the direct determining factor is not the temperature itself, but rather one of its consequences. Tsetse flies and horseflies, for instance, could be carriers of larger numbers of parasites in hotter regions, and the stripes might merely be a defence against these parasites – detectable through the prism of temperature. Second, the black stripes might heat up more than the pale ones, causing slight currents of air between the stripes that cool the animal down. Improbable as this may seem, the internal temperature of a zebra is on average 3°C lower than that of other herbivores of comparable size. The question is therefore whether this difference in temperature is a selection pressure strong enough to explain the appearance and endurance of these markings.

In the current state of knowledge, it's difficult to choose between these last three principal explanations. Indeed, biologists who follow the question closely believe that at present there is no single convincing explanation that can alone account for the zebra's stripes, and that these marks are probably due to a combination of reasons that are difficult to analyse individually. It may be, for example, that stripes first appeared as an effective fly deterrent, and that they were subsequently retained partly for another reason, perhaps the optical illusions they create.

This type of situation can complicate the task of biologists in their attempts to understand the evolutionary history of a characteristic such as the zebra's stripes. Imagine for a moment that stripes appeared on the earliest zebras to provide them with camouflage in an environment that no longer exists, such as a forest of barcodes. Following some sudden climatic event, the forest vanished equally suddenly, but the zebras had already based their social system around their striped appearance, and any individual that was born without stripes would find it very hard to reproduce and hand down its non-striped genome. If selection pressure against stripes was weak because they didn't disadvantage their bearer (while at the same time helping it to disorientate the occasional lion or horsefly), it is conceivable that this characteristic might have been preserved for a function that was different from its original one. In this case, the biologist will be confronted with working out what could have been the initial reason for the adaptation, and discovering the remains of the improbable forest of barcodes.

Getting to the bottom of the zebra's stripes is a scientific adventure. It's a good example of the challenges that evolutionary biologists face on a daily basis, especially those seeking answers to old problems everyone thought solved but new research questions and reopens. Observations in the wild, experiments and computer simulations all play a part in uncovering and explaining evolutionary paths and outcomes. Progress can be slow and sometimes comes from discoveries that are completely unexpected. Who would have thought the zebras' stripes might help them avoid flies, heat, predators and possibly even anti-tank missiles.

PART II

The Mysteries of Animal Behaviour

In order to survive, animals have developed patterns of behaviour that are usually complicated and sometimes bizarre. These can often (but not always) be explained by looking at their evolution, as we shall see in the myriad forms of life in the teeming grasslands of Africa.

Chapter 5

The Air-Conditioning of the Termite Mound

TERMITES ARE VERY SMALL, rarely more than a centimetre long. Termite mounds, on the other hand, may be as much as nine metres tall. Even at an average height of two or three metres, termite mounds are 500 times taller than their inhabitants: the equivalent for us would be living in the Burj Khalifa skyscraper in Dubai, 830 metres high and currently the world's tallest building.

The termites' main tool is their mandibles. Their basic material is a mixture of soil, excrement and saliva. Applied in dry layers, this cement sets and becomes as hard as stone. The mound is crisscrossed by a network of countless tunnels, which can be revealed by means of 'endocasting', so that they can be studied in detail: plaster is poured into the main tunnel (either driving out the termites or killing them on the way), and when it has dried (this can take months) the soil is washed away. The end result is a cast of the interior of the termite mound with its complex structure of chambers and galleries.

The second revelation is that the termite colony, several million strong, does not live in the mound, but instead spends its time in an underground nest, where they breed, cultivate fungus and store food. The colony, consisting of around five kilos of termites and forty of fungus, inhales and exhales as much as an animal the size of a goat. The heat and carbon dioxide produced by respiration have to be evacuated and replaced with fresh oxygenated air. The mound plays a key role in this process. It works as a chimney, a passive regulator of heat and oxygen. In other words, it was termites that invented the first bioclimatic dwelling, some 200 million years ago.

Termite mounds are natural air-conditioning systems, created without the aid of an architect or even of a complex nervous system. Just a joint enterprise by several hundred thousand anonymous individuals. How does it work? For much of the time since the 1960s it was thought termites built mounds that were porous so the wind could rush in and cool it, rather like the natural building ventilation system provided by Persian windcatchers. The idea went like this: with all those tunnels connecting the depths of the mound with the exterior, when the wind blew across the mound on the surface it created a depression which sucked the hot air up from the bottom. In this model, the termite mound cools itself by convection, a phenomenon comparable to what happens when you open the windows in a house.

This really does work: an apartment building in Zimbabwe has been constructed on this principle, and it does indeed consume 90 per cent less energy while saving $700,000 in air conditioning bills annually. But even if this principle does work extremely well for cooling down apartment buildings built by humans, this is not the way that termite mounds function.

Termite mounds are built in two parts: the mound that we see, with its internal network of many large tunnels through which the wind can easily enter, and the nest where the termites live. The galleries in the nest are very narrow, with a diameter of only a few millimetres; they are linked together in a network of small chambers that are full of little holes, like Gruyère cheese. In this very dense 'sponge' air circulation is restricted, hence it follows that the model postulating that termite mounds breathe and are cooled by bulk movements of air is necessarily false, at least in part. Air certainly circulates in the mound, but this circulation doesn't seem to reach the nest.

Another possible analogy for the way termite mounds work is the operation of the lungs. When we breathe in, gas exchange takes place in three phases. In the first air enters the body through the larynx and flows rapidly and in bulk through the trachea. In the third gas exchange takes place not through the bulk flow of air – which the tubes here are too tiny to allow – but through diffusion in the alveoli of the lungs. In between the two the air is subject to the combined influences of the trachea (convective exchange) and the alveoli (diffusion) in a complex process which regulates the link between the two subsystems.

Similar processes may be at work in the termite mound. In the underground maze of sponge-like tunnels just beneath the mound the air circulates by diffusion, while above ground the wider passages allow a large tidal flow of air. Between the two an intermediate system controls the exchange of air from one system to the other that ventilates the nest. Precisely how this

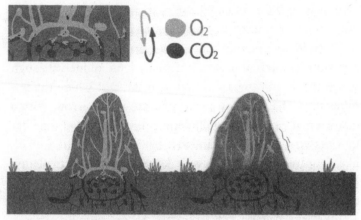

The way in which termite mounds breathe may be explained by
two mechanisms: the pendelluft effect, and acoustic mixing.

system functions is still the subject of research in which two
main avenues are being explored.

The first is that the winds blowing over the termite mound
cause the air to be drawn up through the central channel of the
mound in a chimney effect. As we've seen, this is insufficient
to renew the air in the nest, but it could cause an oscillating
effect in the upper air mass which might be enough to mix the
two layers sufficiently to have a cooling effect on the nest. This
effect in the lungs is known as pendelluft or 'air pendulum'.

The second hypothesis proposes that the numerous tubes
of the termite mound are like a pipe organ. When the wind
rushes into them, this musical theory goes, it makes some of
the tubes vibrate at very low, infrasonic frequencies. You can
conduct an experiment to demonstrate this at home: if you put
some smoke in a bottle and then seal it, after a while the smoke
will sink to the bottom. If you then subject the bottle to low-

frequency vibration, the two layers may mix quite rapidly. This phenomenon is known as acoustic mixing and, if it were shown to be in operation, would be enough to mix the layers of air in the tunnels of the mound.

So maybe termites use the mound like an organ to breathe. The effects, whatever the processes involved, are extremely efficient and one reason for studying them is as a means of improving ventilation in deep mines.

Termites are excellent architects and brilliant engineers. They are also good farmers, cultivating a fungus of the genus *Lepiota*, which they feed with pre-masticated leftover vegetable matter. The fungus then sets about carrying out the delicate stages involved in digesting cellulose, which many termites are incapable of. The termites then consume the sugars supplied by their symbiotic fungus crop. Even this isn't the end of their talents: termites are a keystone species, unique and essential to the ecosystem of the savannah. Their mounds accelerate the formation of soil, increase the size and number of the fruits borne by nearby trees, improve the fertility of insects that live in the vicinity, and generally boost primary productivity all around them. The effects can be seen from space.

Chapter 6

The Impala's Mexican Waves

THE SIGHT OF ANTELOPES GRAZING may seem unre-
markable. And yet behind the repetitive behaviour of
these members of the Bovidae family hides a wonder
of evolution, an example of a form of group behaviour that
shares its mechanics with the spectacular murmurations of
starlings, the synchrony of human applause at a concert, the
choreographed evasion of predators by schools of fish, the
simultaneous flashing of fireflies and even the frustrations of
traffic jams on busy roads.

Antelopes adapt their behaviour to maximise their chances
of survival. The first strategy they adopt is to group together.
Living in herds allows them to reduce the probability of getting
eaten, dividing the odds by the number of other antelopes
nearby, in the effect known as risk dilution. If an animal has
ten neighbours around it, it is ten times less likely to be the
target of the next lion attack than if it were on its own. The

Vigilance is transmitted from one animal
to the next at the waterhole.

other major advantage offered by a group is the multiplication
of the number of eyes available: the more individuals there are
on the watch for hungry predators, the greater the chance of
avoiding any surprise attacks. Communal vigilance therefore
aids the survival of the individual. These two effects are pretty
much enough to justify the interest of group living if you are
an antelope: every time you double the number of friends and
family grazing alongside you, you greatly reduce the chances
of becoming a lion's lunch.

If antelopes were even cleverer, however, they could coordinate their efforts to reduce their exertions: 'You keep watch
and I'll eat, then in ten minutes we'll swap, OK?' In this way,
you would need only one vigilant individual to ensure that all
the rest could ruminate in peace. But antelopes are like sheep.
Instead of having one individual volunteering to go on guard
duty for the common good, they use the strategy of mimicking

their neighbour: 'If my neighbour raises his head to look around, I'll have to raise mine in case he's seen something important. If he puts his head down again, that means that he thinks there's no obvious danger, so I can start eating again too.' Each individual is torn between the need to eat and the urge to copy their neighbour. This simple copying from one individual to the next triggers waves of vigilance, just like a Mexican wave in the stands at a football match. The appearance of these waves of collective vigilance is one of many examples where complex group behaviours are in fact generated by a very simple rule (in this case, copying), repeated many times at a local level (in this case, between neighbours).

How can a behaviour pattern such as this be selected at the level of the individual? You have to remember that what matters in evolution is not to run the fastest, but to run faster than your neighbour. Predators are more likely to catch individuals that let down their guard than they are to catch those who are on the alert. If you are a vigilant individual surrounded by slackers, they will get eaten before you. In other words, being the last to notice an imminent attack puts you in a vulnerable position, and a good way to avoid bringing up the rear is therefore to copy your neighbours as soon as they show signs of vigilance. These complex phenomena derived from simple rules are examples of self-organisation: by looking at the basic principles that motivate each individual, we can understand the complex end result within the group.

Schooling fish are another celebrated example of self-organisation. Fish can group together in shoals of thousands, of all ages and both sexes. Yet despite this variability in their composition, they manage to form groups that are so homogeneous and closely synchronised that you could easily think they were some kind of super-organism with its own nervous system. Despite their striking ability to swirl away from an incoming predator, the rules that operate within a school of fish are scarcely more complex than those governing the vigilance of antelopes – even though in this case the movements to be synchronised are in three dimensions.

The recipe for obtaining a highly organised school of fish is simple and requires only three ingredients. The first can be called close-quarters repulsion. If my neighbour invades my space, I'll move appropriately so that we keep the same distance apart. The second is long-distance attraction. If I am too far away from the group, I move in closer again. I must not get cut off from the rest, and I must not get gobbled up by a predator either. These two behaviours are essential to the cohesion of the shoal as a whole. And the third ingredient is copying my neighbour: whichever direction he goes in, I go too. It is this mimicking of alignment among individuals that makes it possible for all the fish in the school to suddenly and very rapidly change direction as one. The movement travels through the group at the speed of the reflexes of each individual fish.

To demonstrate beyond doubt that these complex phenomena can be explained by a multitude of small and simple rules, researchers have simulated virtual fish, moving them to an algorithm consisting of these three rules. And the virtual fish duly formed homogeneous schools, able to react very rapidly to attacks by a virtual predator. Using this model of three basic actions, the researchers were able to explain numerous examples of group behaviour, from schools of fish to murmurations of starlings, and even crowds of humans.

This model has given us a better understanding of the way crowds respond in panic situations, such as football supporters fleeing a stadium fire. In this sort of situation, it turns out, people respond on an individual level to a few extremely simple rules: avoid other people, get away from walls, run as fast as you can. Thanks to simulations, the researchers have been able to propose more efficient systems of evacuation that can avoid potentially catastrophic situations such as blocked exits.

Groups of people typically express their approval by clapping their hands together, frenetically and/or rhythmically. The more noise, the greater the appreciation and approval, and so an appreciative audience claps their hands as fast as possible. But we may also want to clap in time with other people (perhaps to get an encore), as the collective phenomenon produced by the rhythmic clapping of a whole crowd of spectators is intoxicating in its effect. Two simple rules explain the cycles of noisy unstructured clapping and coordinated slower clapping we often hear at concerts, when many in the audience may be torn between wanting to make a lot of noise and clapping in time with their neighbours. The two modes of applause can compete with each other to produce successive phases in which coordination between neighbours produces waves of applause on a scale matching the size of the hall. As we've seen, antelopes also display two mutually exclusive patterns of behaviour, eating or maintaining vigilance, and they also copy their neighbours. The mechanisms seem to be much the same.

These examples lead to a number of general conclusions. Complex group behaviour isn't the result of complexity at the level of the individual: in spectacular phenomena such as murmurations of starlings each individual follows three simple rules

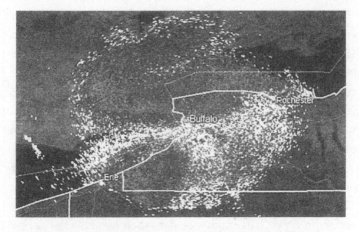

A weather radar image from western New York State
reveals a flock of birds measuring 240km in diameter.

which neighbouring individuals repeat, and so on and on. It
seems to be the case that any group of closely linked members
can display a similar dynamic. This type of phenomenon, where
without conscious planning local interactions combine to form
a coherent system, is known as self-organisation.

For this collective expression to emerge, two conditions
need to be satisfied. The first is the existence of system-specific
thresholds that have to be exceeded: for example in the number
of individuals that interact closely. It appears that there is a critical level below which any collective behaviour will not take off:
for instance, it takes more than one person clapping to set off
general applause in a space the size of a concert hall – which
is why warm-up artists are employed to create a critical mass
of clapping so that a performance does not fall flat on its face.
For shoals of fish, the thresholds are spatial: there is a zone of

repulsion, within which an individual will distance itself from its neighbours, and a zone of attraction, beyond which it will be drawn to them. The second requirement involves what are known as feedback loops. Positive feedback is essential in the development of self-organised phenomena. The more antelopes looking vigilant there are around a given antelope, the more vigilant that antelope will be. The more fish there are swimming in a given direction, the quicker the rest of the group will follow suit. The more people there are applauding, the greater the number of other individuals who will also start to clap. This very simple mechanism, based on a combination of threshold effects, offers an elegant explanation for the formation of waves of vigilance: when enough of my neighbours are displaying vigilance behaviour, I'll copy them and be vigilant in my turn. When the number of these neighbours falls below a certain threshold, I'll return to placidly grazing the savannah. And when this mechanism spreads from one animal to its neighbours, tsunamis of vigilance will form with great rapidity, as these phenomena accelerate with the number of individuals taking part. And of course there is one more prerequisite for the fabulous phenomena to take place: the individuals concerned must copy each other blindly, or, to put it another way, their motivations must be broadly similar.

To sum up, the recipe for self-organised collective behaviour goes as follows. First, a large number of individuals must copy each other blindly or be moved by the same motivations. Second, a set of simple rules must define their behaviour with regard to their near neighbours. And lastly, there must be thresholds defining the point at which feedback loops will be triggered. The general idea, therefore, is that apparently complex behaviours are no more than the result of numerous simple interactions. Just as some patterns of human behaviour that look complex can be explained by a handful of simple rules that don't require the existence of a large brain.

Collective behaviours are very simple to produce, especially when the individuals in the group are broadly similar. They are so simple, in fact, that they are everywhere. Here are just a few examples to think about: traffic jams; stock market crashes triggered by copycat sell-off trading; fireflies that coordinate their intermittent flashes; and, of course, the synchronisation of elements in ecosystems or catastrophic shifts – but that is the subject of another chapter.

Chapter 7

Elephant Dictatorship vs Buffalo Democracy

I N THE LAST CHAPTER, we saw how easily complex collective behaviour can be generated: all you have to do is gather a large number of individuals who are motivated by a set of rules and decision thresholds and who copy each other in a repetitive manner. These individuals must have broadly similar needs, so that following the crowd does not generate any major conflict of interest. Otherwise they will have difficulty in coordinating their movements.

This raises another question: how do individuals in a group agree when they have widely differing interests or perceptions? There are two solutions, one at either extreme of the spectrum: despotism, in which a single individual takes decisions for the rest, or a shared consensus, in which each individual plays an equal part in the decision-making process. To understand the different interests of different individuals it may help to go into a bit of theory. Imagine you are a passenger on the *Titanic 2*.

One night, the captain discovers an iceberg in the ship's path and has to make a difficult decision: should he steer to the left of it, or to the right? He is a seaman, certainly, but he is also keen on statistics and a convinced democrat, and he decides to call together a group of 101 passengers, including you, and ask you all to vote on the direction he should steer the ship, to port or to starboard. When you have all voted, he will follow the majority decision, without question. Naturally, each individual passenger is pretty poorly equipped for making this sort of choice, so let us say that he or she will have a 40 per cent chance of getting it wrong. In other words, each passenger might almost just as well toss a coin, which would have a 50 per cent chance of getting it wrong. The captain watches through a haze of alcohol as you wrestle over which way to vote. And there is no conferring with your neighbours, this is a secret vote. It is nerve-racking: you are convinced the ship will smash into the iceberg and it will all be your fault. While you are pondering over the litany of accumulated errors of judgement that litter your wretched existence, the captain explains that if you are in a state it is because you are rubbish at statistics. You struggle to see what this has to do with it; you think it is more a matter now of how good you are at swimming. So he explains:

If every passenger has a 40 per cent chance of getting it wrong, that means he or she also has a 60 per cent chance of getting it right. In this case, it's as if you took a coin that's slightly weighted on one side, so that it has a 10 per cent greater chance of falling one way than it would by pure chance only, and tossed it 101 times. In the end, the number of votes the passengers cast for 'port'

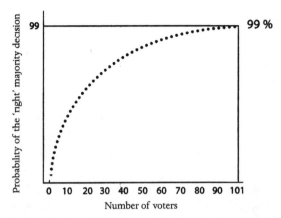

In the *Titanic 2* model, if 101 passengers vote, the 'probability
of the "right" majority decision' being taken (a majority
in this case being 51 passengers) is 99 per cent.

or 'starboard' is in fact described by a binomial law, the
same law that predicts the results of tossing a coin a few
times. All you have to do is apply the binomial formula.
The result is that ninety-nine times out of a hundred, the
majority in a ballot of 101 votes will be right.

In this case, with each person having a 60 per cent probability of being right and with fifty-one voters forming a
majority, you can make a graph, with the number of people
voting plotted on the horizontal axis. This shows that even if
each individual has 'only' a 60 per cent chance of being right, a
majority vote will have a 99 per cent chance of being right. You
end up voting, the majority is right, and you miss the iceberg.
The realisation that under certain conditions a group of people
has more chance of getting it right than a single individual has

Charles Darwin (left) and his less popular half-cousin Francis Galton.

been made on numerous occasions over the course of history. Among those who have made this discovery are Aristotle ('the multitude is the better judge': *Politics*), the Marquis de Condorcet (to whom we owe the analysis illustrated above), and Sir Francis Galton, half-cousin of Charles Darwin.

Galton was a talented polymath, an explorer and a statistician of genius who worked on topics ranging from quantitative genetics to meteorology. He is well known for popularising the sleeping bag (a good thing) and eugenics (not always such a good thing). This fascinating character is often painted as the bad guy to Darwin's good guy, notably because some of his ideas were adopted by Nazi ideologists. The contrast between the two men has been exacerbated yet further by Darwin's quasi-deification as the father of evolutionary biology, a 'liberating' science that offered scientific proof to counter religious dogma about the origins of life and restore humankind to its rightful place in the living world. Galton is portrayed as the man who appeared to use his knowledge to justify speeding up processes of evolution by assisting in the business of selection. Darwin's beard makes him look like Father Christmas. Galton's mutton-chop whiskers seem to characterise him as a Victorian imperialist: it's not surprising he comes out badly in any comparison.

Let us return to the ability of groups to get the right answer, even when they are made up of a random mix of individuals. Francis Galton tested it in a 'guess-the-weight-of-an-ox' game he set up at an agricultural show in Plymouth in 1906. He observed that none of the 800 people who took part got anywhere near the ox's actual weight of 543.4kg but, when he took an average of their guesses, he got a result of 542.95kg, or a mere 450 grams short of the bull's true weight – an almost uncanny degree of accuracy. A century later, in 2007, Scott E. Page published an article demonstrating that for the end result of such a consensus to be accurate, the individuals in the crowd have to express a diversity of opinion. He described his theorem in the following equation:

Collective error = average error *minus* diversity

To reduce the collective error, you can either reduce the error of each individual, or increase the diversity of the individuals (see p. 133). The maths is all very well, but do animals really have a democratic vote, and, if so, how do they cast it?

In animals ballots are carried out by the adoption of a particular posture, gesture or sound. For example, when bees establish a new colony they carry out a dance indicating the direction and distance of the site that has been voted for. Buffalo herds show a similar effect when they move off to a new site:

Buffalos 'vote' by positioning their bodies according
to the direction they want to go in.

when a female buffalo wants to show her preference, she stands
up and positions her body facing in the direction of the spot she
favours, makes a show of raising her head, and then lies down
again. Several other females 'vote' in the same way for their
preferred direction, and the herd will then set off in the direction that is the average of all the individual votes, calculating
the angles with tremendous precision to within plus or minus
three degrees of the mean vote). A fine example of democracy
in action. In cases where two separate directions are chosen, the
herd splits in half.

In different animal species there are all sorts of votes: a
straight majority (red deer and gorillas), an average vote (buffalos), and a quorate decision by a key minority (bees). The

theory behind collective decisions posits that gaining a consensus is one of the most effective ways of maximising the accuracy of the end decision by means of 'collective wisdom', and of minimising the risk of extreme decisions. In theory, no group should accept a despot unless there is a very large discrepancy between the information possessed by the despot and that possessed by the rest, and it is only in this situation that 'following the leader' becomes the most favourable option and therefore the one selected by evolution.

Except that in practice there are many species that cheerfully opt for despotism. In fact, the theoretical models predicting the wisdom of the group assume a whole raft of conditions in order for them to work, and there's the rub. The first of these is the hardest to fulfil: in order for the wisdom of crowds to come into play and for probabilities to accumulate to produce the correct result, as in the *Titanic 2* example, individuals must make their choices completely independently. If people can talk to each other they can change each other's minds, and so put paid to any independence in their opinion; on the contrary, they may form a mishmash of opinions that will stand in the way of the statistical effect predicted by the Marquis de Condorcet and confirmed by Francis Galton. And there are many situations in the animal kingdom where individuals copy each other closely. We know, for example, that their social milieu has a strong influence on the decisions humans make.

In the 1950s the Polish-American psychologist Solomon Asch devised an experiment to demonstrate this phenomenon. He showed a diagram to his students, and asked them to pick the line in the right-hand section that matched the one in the left-hand section, telling them that it was an experiment to do

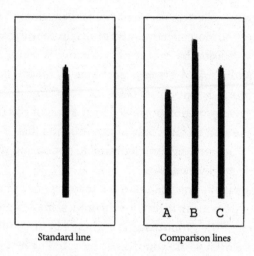

Standard line   Comparison lines

The Asch conformity experiment

with optical perception. When the students were on their own, they gave the right answer, line C, in 99 per cent of cases. But Asch was interested in the results of peer pressure, and did the experiment again with student stooges in which the unsuspecting student who was being tested was unaware that the other students in the room were stooges, instead believing that they were also undergoing the vision test. The planted students and the guinea pig were seated round a table, and each in turn gave their opinion out loud. Asch had asked his actors to give systematically wrong answers. The guinea pig was among the last to give his verdict. The results were clear: three-quarters of the students tested allowed themselves to be influenced at least once, giving the same answer as the others even though they knew it was wrong. (There may be a lesson here for all of us.)

An extreme example of despotism can be found among

African elephants. Female elephants live with their young, sep-
arately from the males, in groups ruled by a dominant female,
the matriarch (photograph 6). She is also usually the oldest in
the group and, since elephants can live to the age of sixty or
seventy in the wild, she is generally a highly experienced individual who knows how best to lead the group through the perils of
the savannah, including predators, famine and drought. When
researchers used loudspeakers to broadcast the sound of lions
roaring to groups of elephants led by matriarchs of different
ages, they found that the older the matriarch, the more quickly
and accurately the herd reacted to simulated danger. A veteran
matriarch was able to tell the difference between the roars of
lions and lionesses, and gave particular attention to the former,
which was not necessarily the case with the younger ones.

In this case despotism has evolved because certain individuals have had time to accrue far more experience than others,
producing a substantial imbalance in knowledge between the
despot and the rest. In short, as the seventeenth-century phi-
losopher Sir Francis Bacon observed, 'knowledge is power'.
Elephants have, however, an unusually long life expectancy
which enables them to accumulate a fund of experience. Their
case does not explain the emergence of despotic hierarchies in
any of the other species in which it is found. Other hypotheses
have been suggested to account for these.

One of them proposes that the leaders are individuals with
the greatest physiological needs. Among zebras, it is the pregnant females who lead the herd to a waterhole, as they are the
ones that need to drink the most frequently. The other members
of the herd follow willy-nilly, as it is in their interests to preserve
the cohesive nature of the group. According to another idea,

leaders are larger, heavier individuals with more aggressive personalities. With their imposing physique, they can dominate competition among males for access to females. They can monopolise reproductive rights (baboons, zebras) and/or may position themselves at the centre of a social network where they can resolve conflicts and improve the group's overall coordination and efficiency, a phenomenon that can be seen in chimpanzees. In these situations submissive individuals suffer the consequences of the inevitable mistakes made by those above them in the hierarchy, but gain the social cohesion that is frequently vital if the perils of life, including predators, are to be survived. Accepting inequalities in power allows the group to compromise between accuracy and cohesion.

In sum, a statistical effect means that a group of autonomous and diverse individuals can offer precise answers when taken together even though any given individual has only the vaguest notion of the best course of action to follow. Mathematics helps explain why democratic processes often seen in the wild generally work. There are also many examples of hierarchical social structures, some occurring when certain individuals are more experienced than others (as among elephants), or are better equipped to resolve conflicts within the group either because they're more powerful (as in chimpanzees) or have the most pressing needs (as in zebras). However, not all despotic regimes among animal species can be explained so easily. The different kinds of social organisation among animals are the subject of much contemporary research in behavioural ecology, some of which may occasionally throw an oblique light on the variety of social and political systems we humans enjoy, aspire to, or put up with.

Chapter 8

The Antelope Art of Sexual Manipulation

MANIPULATING OTHERS TO DO what you want them to do, and manipulating the opposite sex to increase your chances of reproducing with them, are not uniquely human traits. Males of the topi antelope species found in Kenya manage to persuade any females that might be thinking of making a break for freedom to stay quietly at home in their harem. And not only that: they do not just prevent them from leaving, they also take advantage of the opportunity to copulate with them. How?

The answer is simple: when a female threatens to leave his territory, the male raises the alarm with a special snorting sound that he generally uses only to warn of the presence of a predator lurking in the long grass. The male is an accomplished actor, and he gives a convincing performance, striking a pose to signal danger in the direction in which the female is heading, complete with fixed stare, pricked ears and muscles tensed for take-off. The signal could hardly be clearer. The female panics and rushes back to the herd, and in one case out of ten the male will take advantage of this to copulate with her.

The male topi does not bother with any of this if the female is not on heat and therefore is not physiologically ready to provide him with offspring. This is a controlling behaviour aimed at increasing opportunities for sex and the begetting of offspring in a species where sexual competition between males is intense. Sending out false signals in order to get your way is not limited to antelopes and humans: tits, chimpanzees and squirrels raise false alarms to scare off rivals in order to get vital resources for themselves, either food or sexual partners.

PART III

Extraordinary Creatures

If you want to be the life and soul of any social situation, it helps to have a supply of startling biological phenomena up your sleeve. In this section you will find all the necessary ingredients for future social success, from the funny to the suspenseful, and from the poetic to the downright terrifying.

Chapter 9

Dung Beetle Navigation

I F YOU WANT TO PROPEL YOURSELF forwards and straight ahead, rather than going round in circles, it can be helpful to have a fixed point by which to navigate. Celestial bodies are very practical for this, because they are a long way off: even if we move around a lot, our relationship stays more or less the same, and they have made reliable points of reference for sailors since the beginnings of maritime navigation.

Navigating by the sun? Easy. By the stars? Not so easy. In fact, only a handful of species are known to be capable of doing it: common seals, blackcaps, black flycatchers and, of course, humans. But by the Milky Way? Currently only a single species has shown it can pull it off – the dung beetle.

Dung beetles are insects that feed on elephant poo. They cut off pieces of dung, roll them into balls and push them for many metres to their lair. There are some species that are more nocturnal in their habits, and these have developed systems for

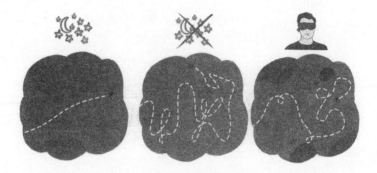

When they have the stars to guide them, dung beetles
crawl in a straight line. In the absence of navigation aids,
they are as disorientated as blindfolded humans. The
dung beetles in this experiment were fitted with tiny
helmets to prevent them from seeing anything.

navigating on moonless nights in order to find their way home.
We know this because in 2013, Swedish and South African
researchers carried out an intriguing experiment in the grasslands of Africa. They covered the heads of some nocturnal
dung beetles with tiny masks, then compared the trajectories
of these 'blind' beetles with others whose vision remained
unimpaired. The results were very clear: the unmasked beetles
headed straight for their destination, while those fitted with
masks went round and round in circles.

The researchers then performed the experiment again. But
this time they did it in a planetarium, where they could simu-
late night skies that were either starry or completely dark, and
with or without the Milky Way; basically, the scientists had fun
creating a range of conditions in order to test how the dung
beetles used light to guide themselves across the savannah

on moonless nights. The results observed in the planetarium matched those in the field: even when they had only the Milky Way to guide them, the dung beetles still managed to crawl in a straight line.

Intriguingly, when deprived of visual cues humans, too, go round in circles. In a 2009 experiment, volunteers were blind-folded and asked to walk in a straight line. Some people ended up walking in circles, some of which were very tight, less than twenty metres in diameter, others wove an erratic course; all of them, however, were quite convinced that they were walking in a straight line. The authors of the research also showed that on moonless nights, even when not blindfolded, their human subjects went round in circles. Which just goes to show that when it comes to navigating by the stars, dung beetles are clev-erer than people.

Chapter 10

Seismic Signalling in the Elephants' Sound-World

People feel so alone and abandoned that they need something solid to cling to, something that will truly last. Dogs are old hat; people need elephants.

Romain Gary, The Roots of Heaven (Les Racines du ciel), 1956

The very word *elephant* conjures up images of these colossal animals plodding in stately fashion through the tall grasses of the savannah. Nothing else looks like an elephant. Their unique bulk and heft lends them a quality that can be almost endearing, their pillar-like legs and chunky bodies may make us smile and their antics with their trunks can come across as comical. But the elephant's imperial bearing, impressive tusks and general attitude of invincibility all command respect. When most people first encounter a fully grown elephant, they are instinctively awed into silence: the sheer presence of this grey colossus, this five imperial tons of flesh, hushes their voices to a whisper.

Elephants are cloaked in myth: their memory is fabled, their intelligence is celebrated, and they are believed to mourn their dead – grouping together in graveyards – and to be frightened of mice. As is often the case, these myths contain both truths and falsehoods (elephants are not afraid of rodents and they do not have graveyards, for starters). Meanwhile, some of the most fascinating things about elephants are all too often passed over in silence – some of these little-known facts deserve to be urban legends.

Since 1993 the English animal behaviourist Karen McComb has worked in Amboseli National Park in Kenya, studying cognition and communication in elephants. Her remarkable findings include not only the adaptive value of age in the leadership of elephant matriarchs, but also the discovery that elephants are capable of distinguishing the ethnicity of the humans with whom they share their living space. It was known that elephants reacted more powerfully and negatively to the clothes worn by Maasai people than to the clothes worn by Kamba people. In an experiment using concealed loudspeakers in the field, McComb played sound recordings of human voices to groups of elephants. The same phrase – 'Look, look over there, a group of elephants is coming' – was said first in the Maasai language and afterwards in the Kamba language. Obviously the elephants could not understand the words, but nonetheless the researchers observed very different responses from them according to the language they were being spoken in. The Maasai people, who raise livestock, regularly come into conflict with elephants and sometimes hunt them. The

Kamba, by contrast, are agriculturalists who have less quarrel with elephants, despite the damage they sometimes do to crops. The fact that the elephants reacted strongly to the phrase when spoken in Maasai demonstrated that they were able to distinguish between the two languages, and that they associated Maasai people with potential danger. When they heard the Maasai recording, they displayed classic responses to the possible presence of a predator: bunching together in a group with tusks pointing outwards; using their trunks to sniff their surroundings, urgently searching for information about the predator; or simply stampeding in the opposite direction from the loudspeakers.

Elephants and their extinct ancestors have suffered hugely at the hands of humans. In the twentieth century alone, hunting and the destruction of their habitat ensured that the number of African elephants fell from around five million to fewer than half a million. In the 1980s, according to WWF estimates, 100,000 African elephants were slaughtered every year for their ivory or their meat. This is not a new phenomenon: recent archaeological research has linked traces of the Clovis people, the earliest human inhabitants of North America, 15,000 years ago, with remains of gomphotheres, members of the pachyderm family that then roamed the Americas. Worldwide, the ancestors of elephants have been hunted for millennia. For 780,000 years, and throughout their range, which covered most of Europe, Africa, Asia and the Americas, the ancestors of elephants died out whenever members of the genus *Homo* settled in these regions. Humans (*Homo sapiens*) played their part in this, of course, but so also did *Homo erectus* and Neanderthal man, *Homo neanderthalensis*. The early elephants were safe only where the human population was too small to hunt them to extinction. They therefore had every reason to associate humans with danger.

In Amboseli, McComb also explored whether elephants could distinguish between men and women. In theory, elephants should react differently to the presence of men, who hunt them, and that of women, who do not. And indeed this turned out to be the case. The next question to investigate was how the elephants managed to tell the difference. The researchers tested the hypothesis that they could tell women's voices from men's by their higher pitch. By adjusting the recordings so that both the men's and the women's voices were pitched at the same level, and then running the experiment again, the scientists obtained results that were startling: even when the pitches of the male and female voices were levelled out to the point where we would generally find them indistinguishable, the elephants could still tell the men's voices from the women's, and reacted negatively to the men.

Karen McComb has shed light on another little-known aspect of the life of elephants: they inhabit a rich and complex soundscape. Their vision is not particularly good, but they have highly attuned senses of smell and hearing. Her team estimated that on average the Amboseli elephants could recognise the calls of fourteen separate family groups, representing a hundred or so individuals. They also showed that the elephants responded to the recorded calls of members of their families that had died since the recordings were made. In a touching anecdote, McComb related how her team had taken the recorded call of a female who had died twenty-three months earlier and played it back to her family. The elephants responded by bunching together, approaching the speaker and calling towards it repeatedly. She notes:

> Playback of the call from the female that had died, to her family unit, elicited contact calling 3 months after her death, and contact calling and approach to the loudspeaker 23 months after her death. Playback of the call from the female that had changed family units, to her original family unit 12 years after the transfer had taken place, elicited contact calling.

Understanding of the acoustic world of elephants has advanced significantly. The Elephant Voices project has, for example, built up a database of the different sounds elephants make, dividing them up into some ten different categories or call types, including a range of rumbles, roars, screams and trumpet blasts. The elephant calls that are audible to humans have a vocal range of over four octaves. However, elephants also communicate by means of sounds that we cannot hear – infrasound.

Elephants can emit very low-frequency rumbles, at between ten and forty hertz, which are transmitted partly through the air and partly through the ground, forming a seismic signal: they can communicate with each other by sending and receiving seismic waves. The ground is a good vector for communication, as in comparison with the air it is relatively little used by other animals, and vibrations that are transmitted through it are subject to relatively little interference. As the ground is denser than air, underground vibrations can travel long distances: the percussive force of a 75kg man jumping up and down has been recorded a kilometre away, while the steps of

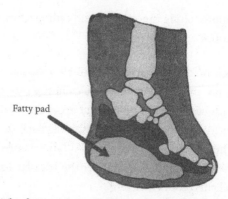

Fatty pad

The fatty pads in elephants' feet help them
to receive seismic messages.

an elephant weighing three tonnes can travel over thirty-six kilometres. Elephants send and receive rumbles through the ground through their front feet, which have fatty pads on their soles to help detect seismic signals and to improve their quality: each cushion acts as a sort of lens for seismic vibrations, and this mechanism enables elephants to increase their sensitivity and to pick up as many signals as possible.

Once received, the vibration is then transmitted up the leg bones to the shoulders, and on to the middle ear. Elephants therefore listen to the ground in the same way as we listen to the air. Among the network of earthquakes that rumble across the savannah, elephants can recognise warning noises from members of their family group, identify where the seismic wave is coming from, and quickly assume a defensive position if necessary.

Elephants are an enigma, but over the past decade some of their mystery has gradually been revealed through the painstaking work of a handful of researchers. We have barely begun to understand the complexities of their social networks, the range of their modes of expression and the rich depths of their cognition, and it is safe to assume that future research will bring yet more startling revelations. Through his protagonist Morel, Romain Gary makes the plea that 'the protection of a margin of humanity broad and generous enough to encompass even the giant pachyderms [may] be the only cause worthy of a civilisation'. He has a point.

Chapter 11

Honey Badger - Weapon of Mass Destruction

I SHOULD START WITH a declaration of interest. I love honey badgers (photo 18). They are amazing animals, the most violent and badass creatures in the biosphere on the planet. They don't just chew things up, they rip them to shreds. They don't emit a call, they produce a piercing scream. The honey badger, otherwise known as the ratel, is a vengeful creature, and other predators are well aware that they need to know their place in its presence, as the honey badger will give them one chance and once chance only. It has given its name to the Ratel infantry fighting vehicle used by the South African army.

For the common good, and for educational purposes, everyone should keep a honey badger as a pet – although with a honey badger *you* would be the pet. Spreading terror among its enemies and aiming for their most vulnerable parts, it attacks on all fronts like a frenzied machine. Honey badgers are also alleged to have the power to heal the sick: a man need only lay eyes on one, it has been claimed, to be restored to full vigour and virility.

Dear readers, I give you the honey badger.

The honey badger (*Mellivora capensis*) is a small member of the Mustelidae family, which also includes weasels and otters, and is found in regions stretching from Africa to India. It is fairly unimpressive to look at: the largest males measure at most a metre in length (tail included) and weigh no more than 15kg – not much more than a poodle. Hardly enough, you might think, to be the scourge of the grasslands, randomly terrorising, impaling and slaughtering other animals at random, and pillaging their prey, offspring and food.

If Genghis Khan had ordered some evil genie to conjure up a genetically modified attack pet that was a sadistic hybrid of a great white shark, a grizzly bear and a giant squid, it would have nothing on the honey badger. Since 2002 the honey badger has regularly won the accolade of 'the world's most fearless creature' in the *Guinness Book of Records*. It has been spotted stealing prey from leopards, fighting with the most venomous snakes in Africa and attacking an elephant – all five tons of it – by its trunk. No problem.

When attacking large animals, honey badgers are reputed to aim for the scrotum. In 1947, James Stevenson-Hamilton, the first warden of the Kruger National Park in South Africa, reportedly observed a honey badger castrating a male buffalo (this deserves a moment's pause: a male buffalo consists of 900kg of aggressive muscle with vicious horns at one end); the account goes on: 'Castrations by honey badgers have been recorded in other animals, including gnus, waterbucks, kudus, zebras and man.' The honey badger aims at the testicles and waits for the poor beast to bleed to death. It can take this cavalier attitude to apparently uneven odds because it boasts a range of adaptations. These include:

- a thick, rubbery skin
- a resistance to snake venom
- a highly developed intelligence
- a reversible anal pouch
- staggering levels of aggression.

The honey badger certainly has a thick skin, including a leathery half-centimetre of it around its neck, said to be impervious to spears and arrows. Its skin is extremely loose so that if an animal grabs the honey badger by the neck in an attempt to immobilise it in a fight, it can twist right round to direct its full fury right at its hapless adversary. Imagine you are a leopard, and a honey badger tries to steal your dinner. You sink your teeth into its neck. It then swivels right round and shreds your face.

Honey badgers have been observed hunting and eating cobras, black mambas and puff adders. Puff adders are the deadliest of all African snakes. Yet a male honey badger has been filmed attacking a puff adder and being bitten by it (and thus injected with its deadly venom). The honey badger then rips the snake to death with its teeth before keeling over – only to regain consciousness a few hours later and tuck into its hard-won meal, looking none the worse for wear. Biologists are trying to understand the molecular mechanism honey badgers use to protect themselves from snake venom.

Honey badgers are clever. In Ranthambore National Park

in the Indian state of Rajasthan, one was filmed trying to catch a kingfisher chick that was trapped high up and out of reach: it rolled a log to a point where it could stand on it to reach the chick. This use of an improvised tool signals a developed intelligence.

There may be quid pro quo arrangements of mutual convenience between the honey badger and several bird species. There are many accounts – none verified, it has to be said – of the greater honeyguide (*Indicator indicator*) guiding honey badgers to bee colonies, where they find the larvae on which they love to gorge. It's also claimed, perhaps more reliably, that some birds follow honey badgers to scavenge what's left behind after they have destroyed the bee colony: the pale chanting goshawk (*Melierax canorus*), for instance, has been seen picking off small reptiles as they flee the micro-earthquakes set off by honey badgers as they dig their way to the bee larvae.

In addition to its arsenal of long claws, sharp teeth and a tank-like profile designed for head-butting, the honey badger also possesses another weapon: a reversible anal pouch that when deployed gives off a suffocating, corrosive stench. This scent gland (a deceptively anodyne term in this case) has been said to help to neutralise the angry African bees when the unwelcome intruder smashes its way into their colonies to gobble up their young. The honey badger is also one of the few mammals capable of running backwards.

In India honey badgers have been seen digging up recently buried human corpses to eat them. In Basra in Iraq in 2007, according to a BBC report: 'Word spread among the populace that UK troops had introduced strange man-eating, bear-like beasts into the area to sow panic. But several of the creatures, caught and killed by local farmers, have been identified by experts as honey badgers.'

The honey badger's reputation must have gone before it.

Chapter 12

The Truth about the Lion King

W HEN *THE LION KING* hit cinema screens in 1994, it moved an entire generation of children to tears at the death of Simba's father Mufasa. It also brought together in the public's imagination two species unknown to each other in the wild – meerkats and warthogs – and made it clear that 'slobbering hyenas' are just plain bad. A scientific analysis of *The Lion King* suggests a degree of scepticism may sometimes be warranted about its portrayal of animal life in the African savannah.

Is Rafiki lost?

King Mufasa's trusted friend and adviser, the 'wise old mystic' Rafiki, is a cheery primate. Rafiki anoints the newborn Simba and in the middle of the film explains to the young lion that his father, although physically dead, lives on in his son. Rafiki lives in a large baobab tree, and is referred to as 'an old baboon'. Yet his colourful face markings show that he is clearly a mandrill (*Mandrillus sphinx*), a species that lives largely in the tropical rainforests of equatorial Africa. *The Lion King* takes place in the grasslands of Kenya on the other side of the continent.

Are lions with black manes more aggressive?

Possibly. Mufasa's brother, Scar, has a black mane. He is also scrawny and treacherous, and is plotting with the hyenas to usurp Mufasa's throne. Scar is the embodiment of perfidy and aggression; Mufasa personifies peace and wisdom. Recent research has confirmed a link between coat colour and behaviour: in some species darker individuals are more aggressive and in lions mane colour can indicate levels of testosterone and aggression. We await studies to show whether males with dark manes are also crooked and devious.

Do hyenas spend their time laughing?

The hyenas in *The Lion King* – and especially one of the principal trio – are perpetually giggling and cackling away. It is a common misconception that hyenas are constantly laughing their distinctive laugh. The spotted hyena is, in fact, a highly social species with a complex language composed of many different calls. The dozen or so different calls that have been identified are used in a variety of contexts: as a warning of an imminent attack, in encounters with other hyenas, as a sign of submission, and so on. The 'chuckle' is just one of these calls. As a response to aggression from another individual, it basically means, 'Leave me alone!' Not really a laugh then.

Are lions hunters and hyenas scavengers?

Scar pulls off a coup d'état and unleashes a reign of terror in which the lionesses are forced to hunt in order to provide food for the lazy and ravenous hyenas. In fact, spotted hyenas are excellent hunters, and most of the meat they consume is from prey they have hunted themselves. Scar is right when he says that it is the lionesses' job to do the hunting. But the kings of the jungle are more than happy to feast on carrion, and will push other predators off their kill in order to steal it from them. Actually, lions are bigger scavengers than hyenas.

Do elephants have a graveyard?

Young Simba leaves his father's kingdom to explore the mysterious 'shadowy place' on which (inexplicably) the rays of the morning sun never fall. This area turns out to be an 'elephants' graveyard', a sort of wasteland-cum-tip where menacing junkie hyenas rule (and threaten to bite chunks out of our gentle heroes). So far so good, except that elephants don't have graveyards. This myth has probably arisen from two observations. Elephants, like humans, are attentive to their dead. They often linger over their bodies: there are many stories of elephants spending several days beside other elephants' corpses, and even covering them with branches, The bones of several elephants are sometimes all found in the same place, as though they had gone to a single spot to die there together.

The reality is simpler: elderly elephants tend to die near waterholes where they are drawn to soothe their aged bodies with the water and to feed on the water plants that are softer for their toothless jaws to masticate. The old elephants can get stuck in the mud around the waterhole and there, either because they are unable to free themselves or because they are so weak that they lose consciousness, they die.

Can animals be superstitious?

A significant consequence of Scar's despotic regime is that the whole environment becomes out of joint: the grasslands, so lush at the beginning of the film, are transformed into a sandy desert, the sky turns grey, herbivores die of hunger and, as one of the hyenas complains, there is nothing left to eat.

The animals in the film must have short memories: they have forgotten that the dry season comes every year and that during the dry season the grasslands become parched. Some countries in sub-Saharan Africa experience no rainfall at all for six or eight months of the year. Plants dry up and wither to reveal the bare soil (which in many southern African grasslands is sand), and may then spontaneously ignite, filling the air with smoke and turning the sky a distinctive purplish grey. In a particularly fierce dry season, some herbivores may perish. In short, the phenomena of 'ruin and devastation' that are attributed to Scar's despotic regime are just part of the annual cycle of the seasons of which animals in the wild tend, for obvious reasons, to be keenly aware.

Is there incest in the royal family?

As a cub, Simba has a best friend called Nala. After his father's death, Simba is told to leave the kingdom by Scar, and grows up outside the homeland. The years go by, and then – by a remarkable coincidence – he comes face to face with a lioness who turns out to be Nala, who has fled the devastated kingdom of the lions. The two cubs are all grown up now, and the hormones coursing through their systems work their magic: they nuzzle, play with each other's manes, gambol and play-fight while the rest of nature throbs and pulses around them. United, the happy pair will make a stand against Scar's evil empire, and bring down their treacherous dark-maned uncle, and replace his despotic regime with the true circle of life and with their own dynasty of little blond cubs. So they all live happily ever after. But wait – what if those little cubs happen to be the off-spring of blood relations?

Shock revelation: Nala and Simba have to be related to each other. At best they are cousins. At worst they are half-brother and half-sister. This is the way a pride of lions works: there may be one dominant male or a coalition of two, and these males monopolise the reproductive rights of many females. The other males are either too young to rival them, or are expelled from the pride's territory. In the case of a coalition, two brothers will form an alliance to force another dominant male out of his territory, and will then share his reproductive rights between them. This is the most logical explanation for Scar's presence in the pride: he and Mufasa must have established a coalition. Yet Scar does not seem to be particularly fulfilled sexually, as Mufasa appears to monopolise the females (and Scar feels understandably aggrieved). There is, therefore, good reason to suppose that Mufasa has fathered all the cubs, Nala and Simba included. So Nala and Simba are most probably half-brother and half-sister, and their family tree risks containing record levels of inbreeding, and of producing baby lions displaying a wide range of recessive disorders and birth defects.

The Crater lions of the Ngorongoro Conservation Area are a good illustration of this effect. The Ngorongoro Crater, the largest extinct volcano in the world, supports an isolated population of closely related lions, recently reconstituted from a mere handful of individuals and forming what is known as a genetic or population bottleneck. The effects of this inbreeding can be seen in the males, with nearly half of their sperm displaying abnormalities. Behold the future that awaits Simba and Nala's offspring.

PART IV

The Human Factor

Humans systematically come face to face with other species in the biosphere: sometimes we act as diabolical agents of destruction, and sometimes – if not all that often – as Promethean and quasi-divine benefactors. What interests us here are the relationships and interactions between this naked ape and the other species that inhabit the grasslands, and the extent to which any perceived gulf between them is in fact a mirage.

1. A red-beaked oxpecker of the Buphagidae family prepares to land on a giraffe's neck. These birds feed on the parasites that live in the giraffe's coat, which is good for the giraffe, but they also peck at the animals' skin to open up wounds and drink their blood, which is less good – making an excellent example of the often fragile boundary between symbiosis and parasitism.

2. Female spotted hyenas are difficult to distinguish from the males, even though they are generally larger. Females share numerous anatomical features with the males, among them a clitoris that may be 18cm long and that to the naked eye looks very like the male penis. Inserting a penis into a pseudo-penis is not the easiest of moves, hence the suggestion that by making copulation so difficult the pseudo-penis may enable the female to choose her mate. Nevertheless, the evolutionary reasons behind this organ remain a mystery to biologists. © Wikimedia Commons, *Stig Nygaard*

3. A group of blue wildebeest (*Connochaetes taurinus*) heading for a new feeding site. In the late nineteenth century, the accidental introduction of the rinderpest virus to East Africa caused high mortality, not only among domesticated cattle but also among wild ungulates such as buffalo and wildebeest. Within the space of a few years, the morbillivirus spread throughout the African continent. The number of animals it killed remains difficult to estimate, but can probably be counted in the millions.

4. A female impala looks on while two males spar. The breeding season is over for these males, however: this is just a play fight, with no territorial rights at stake.

5. An impressive herd of buffalo gathers at the Ngweshla waterhole in Hwange National Park, Zimbabwe. Buffalo move around in large groups, reducing predation risk through numbers and group vigilance.

6. Elephant herds, each led by a matriarch, arrive en masse at a Nyamandlovu waterhole, which is well known for the large numbers of these animals that it attracts. The age of the matriarch is an important factor in ensuring the safety of the herd, as the long experience of the older females enables them to react more effectively at times of danger.

7. At nightfall, giraffes move closer to the dwellings on the fringes of Hwange National Park to seek safety from their predators, which – with the occasional exception of bold lions – do not dare to venture among the human habitations.

8. The tree stump poking out of the top of this abandoned termite mound indicates that the mound was an important source of fertility when it was active. The complex internal architecture and solutions devised by termites to ventilate their mounds make this heap of mud worthy of a second glance.

9. It is rare to spot an active leopard in daytime. I scarcely had time to grab my camera before he was loping off through the long grass, probably to find a spot to sleep until nightfall, his preferred hunting time.

10. Female kudus (*Tragelaphus strepsiceros*) come to drink at sunset. Vigilance is essential at this time of day, when predators are most active.

11. Whether narrow or wide, zebras' stripes vary greatly between individuals. With a little practice, it is possible to distinguish between them with the naked eye. The white-on-black stripes of the animals' right and left sides display absolutely no symmetry. There have been many models attempting to explain the mechanisms by which zebras got their stripes. One of the earliest theories was put forward by Alan Turing, in 'The Chemical Basis of Morphogenesis'(1952). Turing proposed that patterns could arise from numerous local interactions, in a process described as a 'reaction-diffusion' theory. His ideas would help to explain many biological phenomena.

12. Many have suggested that zebras' stripes are for purposes of recognition, but the hypotheses that are now under serious consideration are altogether more exotic.

13. This foal, only a few months old, still displays the brown colouring of newborn zebras.

14. This young male kudu keeps a vigilant eye on his surroundings at sunset: a perilous time of day for the herbivores of the savannah, when many predators such as lions start to hunt.

15. Female impalas. These exquisite creatures live in herds of several dozen individuals dominated by one male.

16. If his father were to be usurped by another male, this young lion cub in the Serengeti would be in mortal danger. When a dominant male is overthrown, the new arrival systematically kills his cubs. © Flickr, *ganesh raghunathan*

17. It may weigh a mere 10kg or so and be under a metre long, but the honey badger is a pint-sized distillation of aggression and remarkable powers. It can run backwards, survive the most venomous snakebites, will attack buffalos without fear, and often aims for the testicles. © *Anthony Bannister/Photoshot*

Chapter 13

How to Turn a Lion into a Cub-Killer

H UMANS EXERCISE A MAJOR INFLUENCE on the biosphere. So widespread has human domination of biological and geological processes become that many scientists refer to the current geological epoch as the Anthropocene (from the Greek for man, *anthropos*). The spread of *Homo sapiens* across every landmass on the planet has frequently coincided with the disappearance of numerous indigenous species, and the rate of species extinction globally is on a par with other major biological crises in the geological past. Human activity has had an impact on biodiversity in many other ways, not all of them negative. Humans can also influence animal behaviour, sometimes in the most surprising manner.

Let us start with a simple example from the grasslands: the great migrations. Many of Africa's grassland ecosystems are strongly influenced by the seasons, with rainy season

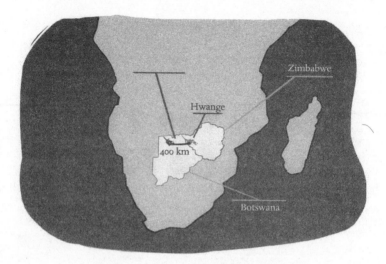

The traditional elephant migration route from
Hwange National Park in Zimbabwe to the
Okavango Delta in Botswana covered 400km.

following dry season and conditions veering from extremely
wet to tinder-dry. In the dry season there may be no rainfall
whatsoever for months on end, and riverbeds and waterholes
may dry up completely. The animals therefore do the obvious
thing and migrate, heading off in search of water elsewhere. In
Hwange National Park, where I worked, the elephants would
historically trek as far as the Okavango Delta in Botswana, or
the Zambezi on the Zambian border, in their search for water.
In both cases, their next trunkful of fresh water lay several
hundred kilometres away.

In the early twentieth century, the first African national
parks were established. The animals within their confines were
now protected from illegal poaching, and their populations

were allowed to flourish. In the Hwange, another measure was quickly adopted. The park had no natural surface water: to remedy this, boreholes were drilled in order to pump water up from deep underground, so that the man-made waterholes were kept full of water even at the height of the dry season. As a result, the animals in the park no longer need to migrate to find water; now they have as much as they want *in situ*. By changing one of the parameters of the habitat, humans have ended migration among the animals. In the Hwange, this has led to an appreciable growth in the elephant population: now baby elephants no longer die of thirst on the perilous journey to the Okavango Delta, and the local population has probably increased to over 40,000. Throughout the dry season the diesel pumps hum away, bringing water up to the surface for animals that have become a settled population.

We do not generally set out deliberately to alter animal behaviour but it can happen as an unintended consequence of our actions. A striking example can be seen in the opportunistic behaviour of the North American mammals – raccoons, bears, foxes and the like – that are attracted to human settlements to scavenge for discarded food and raid dustbins. This happens all over the world; in the Hwange we even had a troop of chacma baboons (*Papio ursinus*) that specialised in sorting and recycling our rubbish.

Of all the vertebrates it is birds that offer the most impressive examples of adapting their behaviour to our technological civilisation. In the early part of the twentieth century, milkmen in Britain were faced with a problem: by the time their customers took in the bottles of milk they left on their doorsteps, which were then uncovered, the cream on top of the milk would often have vanished. The culprits were blue tits (*Cyanistes caeruleus*), which simply came to drink their fill from the bottles. Within a few years, milk delivery companies added aluminium tops to the bottles to keep the milk fresh. But they had not reckoned with the intelligence of the blue tits: a few individuals worked out how to pierce the tops in order to get at the cream, and within a very short time their neighbours had copied them, and the technique had spread throughout the British blue tit population. By the early 1950s, the entire blue tit population of a million or so individuals knew how to get the cream off the top of the milk. Then tastes changed, customers began to opt for semi-skimmed or skimmed milk, rather than the traditional full cream 'gold top', and fewer people had milk delivered to their door. The blue tit behaviour has now died out.

Many bird species incorporate coloured plastic detritus when building their nests (photo 19), using the bright colours as signals to neighbouring birds or to attract females. Among territorial birds such as the black kite (*Milvus migrans*), the largest individuals with the best territories use more plastic in their nests than weaker individuals with less desirable territories. This eye-catching display reduces conflict: each individual can size up a rival's strength at a glance, and so avoid getting into fights in which they are likely to be beaten. Any weak bird that tried to cheat by adding plastic to its nest would run the risk of getting into a potentially lethal fight with a stronger bird. The plastic detritus serves as a visible indicator of an individual's status, so reducing territorial conflicts.

House sparrows (*Passer domesticus*) incorporate cigarette ends in their nests, where the residual tobacco acts as a powerful insecticide to protect their chicks from parasites. Birds in the wild use sprigs of aromatic plants such as thyme in their nests as an antiseptic and a deterrent against parasites, and it appears that cigarette ends make an effective urban substitute. The tobacco plant originally used nicotine as a protection against leaf-eating parasites: when sparrows use cigarette butts in their nests they're exploiting nicotine's original function. Better still, the BBC filmed crows in a Japanese city dropping tough-shelled nuts from overhead wires on to a busy road, so that passing cars would crack them open as they drove over them. The birds soon worked out that if they did the same thing on a pedestrian crossing and waited until the light turned red and stopped the traffic, they could collect the nuts at their leisure without running the risk of being run over.

A recent study suggests that birds feeding on roads somehow know their speed limits. Researchers from Quebec drove at different speeds on a variety of roads in western France. Whenever they encountered birds on the road, they measured the distance between their vehicle and the bird at the point at which it decided to fly off. The results were startling: the distance was not related to the speed of the car, but rather to the speed limit in force on the road in question. On a road with a 90kph limit, for example, their research showed that birds will generally fly off when a vehicle is seventy-five metres away from them, and this distance remains constant whether it be a tractor chugging along at low speed or a high-powered car in a hurry travelling at twice the limit. In the former case the birds will take off earlier than they need to; in the latter case they will be much too late, as at that speed the car will cover seventy-five metres in one and half seconds. Does this mean that birds have learned to read road signs? One rather doubts it. Another explanation is that birds learn to work out the average speed of passing vehicles: on any road there will be some drivers who drive a little over the speed limit and others who drive a little under it: their average speed will therefore be close to the speed limit, and it is this average that birds 'know' as they peck away in the middle of the road. Drivers with experience of less intelligent bird behaviour may think more work needs to be done on this intriguing phenomenon.

In 2014, a research team from the National Taiwan University discovered another example of the hijacking of human infrastructures by a wild animal species. Tiny Mientien tree frogs (*Kurixalus idiootocus*), which make loud, long-distance advertisement calls during the mating season, are frequently found congregating in concrete roadside storm drains during this period. For the frogs, the drains act as urban canyons, enhancing the acoustics of their mating call and increasing its range. These little frogs can therefore be added to the list of animals that have learned how to use human structures to their own advantage.

Sometimes the consequences of human actions on animal behaviour are more unexpected, and in certain cases they are unequivocally negative. In the grasslands, the most heinous example of this is found in the link between trophy hunting and infanticide in lions. Lions live in a social system in which one male may monopolise the reproductive activity of five or six females. The position of the older males is constantly being challenged and usurped by younger ones, and a male lion has an average window of two years for reproduction – the time between his becoming strong enough to unseat another male and becoming too old to resist being ousted in his turn. The lionesses are not sexually receptive as long as they are rearing a litter of cubs, so when a young male overthrows an older male in the pride, he also kills all the cubs. Once the females have lost their cubs, they become receptive once more and ready to mate with the new male. Although the lionesses are ferocious in defending their young, it is estimated that a quarter of all lion cubs that die in their first year are killed by an incoming male.

Trophy hunters pay to kill male lions, among other large predators. This is a distinctly unsporting activity, typically involving helicopters or 4x4 vehicles. The larger the lion, the happier the hunter, so for maximum client satisfaction the quarry has ideally to be a large adult male in the prime of life. These are also the individuals that are most likely to have a harem of females with cubs. When the hunter pulls the trigger, he or she unleashes a tragic chain of events, as new males move into the group to take the place of the slaughtered male and kill all his cubs. In June 2015 the crossbow shooting by a trophy hunter of Cecil the lion (who was named after Cecil Rhodes) in Hwange National Park in Kenya caused widespread dismay around the world and has since led the United States to add more sub species of lions to its endangered species list, which makes it harder for US citizens to go and kill them. So perhaps Cecil did not die wholly in vain.

The list of animal behaviours caused or modified by humans is long, and getting longer by the day. From an impartial and objective viewpoint, humans can be considered as a species that exerts a major influence, consuming resources but also making new ones available. It is always useful to remind ourselves that this state of affairs is not entirely new: it is nearly 30,000 years since the first domesticated animal, the dog, was genetically selected by humans, and at least 60,000 years since humans started using forest fires as a way of making hunting easier and changing the biodiversity in a particular area, long before the invention of agriculture – so-called 'fire-stick farming'. By setting more recent events in this context, we may be less tempted to consider all human impact as 'unnatural', and as antagonistic to the 'virgin' state of what we imagine to be pristine, untouched nature. From the beginnings of human history, we have shaped our environment and been shaped by it, modifying the behaviours of the species we share our lives with and being modified by them. The difference today is simply one of scale.

Chapter 14

Catastrophic Change

THE DUST BOWL is the name now given to the period in the 1930s when dust storms ravaged the North American prairies. The storms would blow up suddenly after a severe drought, in an environment that had been farmed intensively for many years. The spread of the tractor and of deep ploughing (combined with a lack of understanding of the local ecology) led to the systematic destruction of the deep roots that had previously stabilised the sandy soils of the arid plains. In the absence of the native grasses that historically maintained levels of humidity and bound the soil together, billions of tonnes of loose silt and sand were picked up and carried away by the fierce summer winds. The effects were felt in winter too: in 1934–5, red snow fell on New York, coloured by dust from the middle of the continent. On 14 April 1935, known as 'Black Sunday', dust storms swept right across the Great Plains from Oklahoma to Texas, whipping up 300 million tons of dust and sand and creating the first ecological refugees in American history. During the 1930s, some two and a half million people who had lost their livelihoods because of these 'black blizzards' abandoned the Great Plains and made their way to the coast to look for work, and especially to California – as described by John Steinbeck in *The Grapes of Wrath*. This desertification shows that major environmental changes can sometimes take place within the space of just a few years.

How can we predict and avoid catastrophic events on such a vast scale? The question is particularly pertinent at a time when desertification is more of a threat than ever: 3.6 billion hectares are currently suffering from desertification around the planet, representing over a third (36 per cent) of its exploitable land. One answer is to look at a specific kind of catastrophic change – the avalanches that occur in a number of the earth's ecosystems.

An avalanche occurs when a body of something (such as snow) slides from a stable state on a mountainside to a different stable state in a valley, under the very slight impulse of an external element, for instance new snowfall. It may snow for three hours, and then all of a sudden one last flurry of flakes will unleash disaster. It should be noted that the impulse required might be tiny: the mass of the final flurry could be negligible in comparison with the mass of snow that will slip down the mountainside, but could push the amount of snow on a slope beyond the stability threshold.

In describing catastrophic changes such as avalanches, scientists use the term resilience, meaning the system's capacity to suffer major perturbations without losing stability. The more resilient the snow, the greater the weight needed to loosen it and set off an avalanche. The speed and intensity of a catastrophic transition may be increased by a positive feedback loop that runs out of control: the bigger the avalanche becomes as it cascades down the mountainside, the greater the quantity of the snow it will pick up; like a rolling snowball, the bigger it gets, the faster it grows. To sum up, to concoct a good avalanche-type catastrophic transition, the necessary ingredients are different stable states, plus an external force that exceeds a certain threshold, plus snowball effects that feed the transition from one state to another.

Desertification occurs in arid environments such as grasslands. As an ecological category, 'grassland' is something of a portmanteau term, encompassing systems ranging from arid to semi-arid and from treeless to densely forested, and including systems where successive dry and rainy seasons occur. In practice it denotes a continuum, and the same area of grassland may gradually move from one state to another, under the effects of climate change, diminishing rainfall or pressure from herbivores or forest fires.

Imagine, for instance, that the rainfall on an area of forested grassland is reduced. Gaps will begin to appear in the tree cover. Reduce the rainfall further, and the gaps will start to join up to make lines, and the lines will join up and form mazes. Very soon, there will be only a few patches of forest left. This gradual transition does not necessarily lead to a catastrophic outcome: if the rainfall increases, the trees will come back again.

If we examine the ecosystem in more detail, we see that the trees encourage the presence of water in their immediate vicinity by providing shade, while their roots reduce evaporation

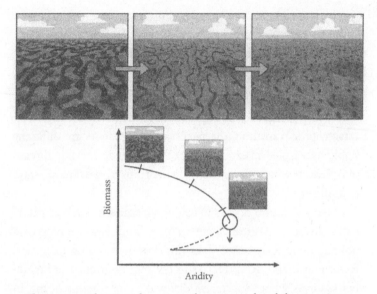

The more arid an area becomes, the more isolated the remaining patches of vegetation are. Every arid ecosystem has a tipping point at which its biomass can be reduced to a catastrophic degree.

and increase local humidity. In other words, when there is very little water available, the only place where trees can grow is next to other trees. This is the reason why, when water is scarce, they form small patches of green in an otherwise bare landscape: they have to live in close proximity in order to survive. In short, the more trees there are, the more trees there are likely to be. Conversely, the bare areas are extremely arid and are eroded by the wind: as soon as there are no more roots to bind the soil, the last nutrients vital for the establishment of new plants are quickly blown away, making it even more difficult for new shoots to grow. In short, the fewer trees there are now, the fewer trees there will tend to be in the future.

When an area covered by trees dips below a critical threshold, it is often observed that the ecosystem suddenly collapses and the trees disappear: this is desertification. Desertification is a catastrophic transition: the transformation of a landscape from patchy trees to desert can be brutally sudden. By contrast, returning in the other direction – that is, reducing the degree of aridity to a level far lower than that at which the transition took place – is a much longer, slower process. There are two feedback loops, one favouring plant growth and the other reducing it. When the level of vegetation dips below a given threshold, the former grassland quickly turns to desert. When the ingredients are all in place, namely:

- two stable states: with vegetation (savannah) and without vegetation (desert),
- external conditions of aridity that exceed a given threshold and
- domino effects that increase the speed of the reaction,

catastrophe lurks around the corner.

Thus when a threshold of intensity is passed, what has been a very gradual process of change can become a reaction that is as rapid and fundamental as it is unexpected and hard to reverse. It is impossible to know in any particular system precisely where this threshold lies. Once the changes are in process they are likely to be irreversible, for it requires far more energy to restore an ecosystem after a transition than to repair it beforehand. The American Dust Bowl created a desert of such stability that more than 200 million trees had to be planted to make the Great Plains once more fit for agriculture. These trees had the effect of boosting local fertility, conserving levels of humidity and binding the soil – in other words, all the positive effects described above. The Sahara offers an example of an even more spectacular transition. Just over 5,000 years ago it was a great plain with forests and lakes. Then changes in the amount of sunlight absorbed by plants and the amount reflected back into space, which had been at work for millennia (and is now known as radiative forcing) may have crossed a threshold, triggering a snowball desertification effect that created the largest hot desert in the world. Exactly why the Sahara should have changed so dramatically is, as yet, uncertain.

Chapter 15

Human Evolution and its Impact

S OME OF THE MOST IMPORTANT EPISODES in our evolutionary history have played out in the grasslands of Africa. One of the most significant was the transition from quadrupedalism to bipedalism, which took place in East Africa some four million years ago. By getting up on two legs, our ancestor Australopithecus could see further, walk for longer and use their hands more freely. The fossil remains of various species of Australopithecus have been found in a strip stretching from the Sahel to South Africa, and palaeontologists have given affectionate nicknames to the most notable individuals: Lucy (*Australopithecus afarensis*), discovered in Ethiopia in 1974, needs no introduction, and there are also Abel (*A. bahrelghazali*), found in Chad in 1995, and Mrs Ples (*A. africanus*), discovered in South Africa in 1947.

Then came the ability to fashion the first stone tools, two and a half million years ago, which coincided with the appearance of *Homo habilis*, the first member of the genus *Homo* to which we also belong. The tools that *Homo habilis* made – consisting initially of simple choppers – enabled them to cut up the grassland animals that they hunted as part of their diet. After this came the domestication of fire, which may have started as early as one and a half million years ago, coinciding roughly with the appearance of an inventive relative, *Homo erectus*. Probably inspired by frequent bush fires, *Homo erectus* mastered this element, taking advantage of the heat, light and protection against predators it provided, and above all enjoying the possibilities it offered for new ways of preparing and eating food. Some scientists believe that this species was the first to cook their food: they base their arguments on the small lumps of clay found in archaeological sites linked to *Homo erectus*, which could only have been produced by a very intense and localised heat, such as a fire burning in a hearth. But there is considerable debate over the exact date at which *Homo erectus* first mastered fire.

Whenever it first developed, the ability to cook had numerous repercussions: cooking softened foods, made the nutrients they contained more easily available and reduced the time needed for chewing and digestion. A morphological effect of this was a reduction in the size of teeth, while the fact that less time was spent chewing food freed up more time and energy for doing other things such as making tools, interacting socially, or even migrating across the landscape. The brain of *Homo erectus* was appreciably bigger than that of their ancestor *Homo habilis* (with a volume of 980cm^3 as opposed to around 600cm^3), and some individuals had brains approaching the size of human brains today (1100cm^3). The cooking of food could well be responsible for this growth in skull size: of all our organs the brain is the one that consumes the greatest proportion of our energy, with almost a fifth going to power brain activity; the increase in resources provided by cooking was able to support the energy requirements of a larger brain. *Homo erectus* made good use of this added brainpower, leaving Africa and its grasslands behind to spread all around the Mediterranean coast and as far as Asia. They even colonised islands that had hitherto been inaccessible, such as Crete in the Mediterranean or Flores in Indonesia, by floating across on driftwood or perhaps building rafts. But whether they stayed in their African cradle or emigrated to Asia, *Homo erectus* still relied heavily on species often associated with grasslands, especially elephants. From the southern tip of Africa to Spain, the numerous archaeological sites associated with *Homo erectus* reveal the remains of elephants, which were butchered for food, while their bones were fashioned into tools.

Our ancestors were members of the *Homo erectus* species who stayed in Africa and did not travel. As time went on, a new species emerged: *Homo sapiens*. The oldest *Homo sapiens* fossils so far discovered were found in the Omo Valley in Ethiopia, and date from 195,000 years ago. However, recent scientific research places the origins of our species considerably earlier, at 340,000 years, almost doubling its age. This recent discovery, made in 2013, came about in a singular fashion: Albert Parry, an Afro-American living in South Carolina, sent a DNA sample to a company called Family Tree DNA for genealogical analysis. When he carried out further testing on the sample, Michael Hammer, a geneticist at the University of Arizona, was surprised to see that the Y-chromosome, carried only by men, was quite different from

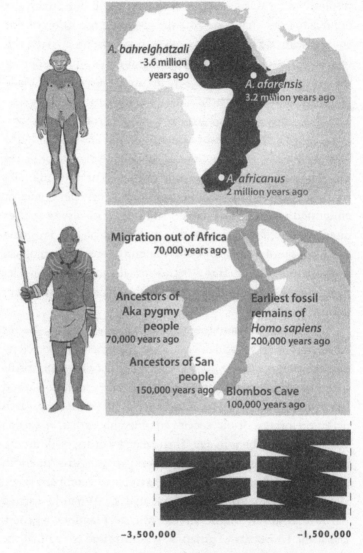

Some key stages in the closely interwoven
history of humans and the savannah.

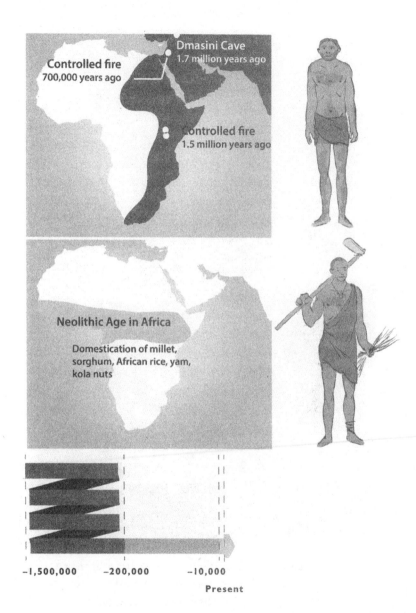

Dmasini Cave
1.7 million years ago

Controlled fire
700,000 years ago

Controlled fire
1.5 million years ago

Neolithic Age in Africa

Domestication of millet,
sorghum, African rice, yam,
kola nuts

−1,500,000   −200,000   −10,000

Present

Humans have long hunted the animals of Africa, as shown
by the rock paintings executed by the Khoisan hunter-
gatherer tribes in the Drakensberg, South Africa.

the Y-chromosomes he was used to seeing. When he calculated the time necessary for evolution to produce the difference he observed between this chromosome and the rest, Hammer came to the realisation that Albert Parry was descended from a lineage of humans whose ancestors had split off from other humans 340,000 years ago. In other words, humanity is older than the fossil records suggest. More detailed investigation revealed that Albert Perry's close forebears probably came from Cameroon, and more specifically from western Cameroon, close to a place where other interesting facts about our biological past have come to light, thanks – once again – to genetics.

The Aka people are a pygmy tribe who form one of the oldest branches of humanity, having split off from other humans some 70,000 years ago, in the area now occupied by Cameroon in Central Africa. At the same time, the human population in the Horn of Africa, the ancestors of all non-African humans, were setting off on their great voyage from Africa and reaching the Arabian peninsula. The pygmy peoples, meanwhile, became specialists in living in Africa's equatorial rainforests, and for a long time it was thought that they remained genetically isolated from other populations. But recent studies show that to this day the genome of these pygmy peoples contains a fragment inherited from other very early human populations that are as yet unknown. At the point when this hybridisation took place, these mystery populations were so different genetically from the pygmies that they may even have constituted a different species of the genus *Homo*. The pygmies must have interbred with this ancient people some 40,000 years ago. Although as yet we have no fossil record of these people, they left their imprint on the pygmies' genome.

The pygmy peoples are not unique in this respect: extensive genetic research and the examination of ancient DNA have uncovered at least two other hybridisation events involving human populations and another species of the genus *Homo*. The event that has been most written about is without doubt the hybridisation between *Homo sapiens* and Neanderthal man: we know now that on average 3 per cent of the genes of non-African humans (whose ancestors left Africa 70,000 years ago) is inherited from Neanderthal man. Whether this means that Neanderthals were technically a different species is still a matter of debate. Such hybridisation events are highly significant, as we know that many genes we inherited from these vanished relatives are now under strong selection, meaning these are useful genes on which we depend for our survival. Some of them are immune genes, others are related to resistance to ultraviolet rays, and others again are linked to keratin, the protein that forms an essential part of skin and hair. It may not be all good, however, as it has recently been suggested that some of these Neanderthal genes may be associated with mental health problems. Finally, up to 6 per cent of the genes in the genome of many populations of South-East Asia and Oceania come from another, deeply mysterious group, known only by the name of the cave in Siberia where a handful of their bones have been found – the Denisovans. As yet we know little of these extinct relatives, but one thing we do know for certain: our history is one of hybridisation.

To return now to the grasslands, which have been home to some human populations since well before humans left Africa: the San people split off from other human populations at least 150,000 years ago, and until the discovery of Albert Parry's

Y-chromosome it was thought for many years that they were the world's most ancient human population. The various nations who make up the San people are hunter-gatherers who live in the arid regions of southern Africa, in particular Botswana. They have an intimate understanding of the arid ecosystems in which they live and are highly skilled and knowledgeable naturalists. When setting off on lengthy hunting expeditions, for instance, they consume parts of the *Hoodia gordonii* plant, which they believe suppresses the appetite, and they dip their arrow heads in a poison called diamphotoxin produced by beetle larvae: this gradually paralyses the muscles of the animals they hunt, chiefly antelopes such as elands and sable buck. Rock paintings found throughout southern Africa and dating from 20,000 years ago show scenes depicting a close relationship between the San peoples and the local fauna, with some painting, probably representing shamanic rituals, even depicting half-human half-animal metamorphoses. Similar scenes were also painted at the other end of the continent, in the middle of what is now the Sahara, where 7,500 years ago a vast area of grassland was grazed by a wide range of herbivores: in the Sahara of Palaeolithic times, buffalo, antelope and giraffe all thrived, as seen from the depictions of them painted in ochre and carved into rocks.

This close relationship between humans and the African grasslands lasted into recorded history, when humans made major changes to these ecosystems. The grasslands of West Africa look to us like natural landscapes, for example, but in fact they have been considerably altered by humans, especially in a strip stretching from West Africa to East Africa, where the earliest forms of agriculture emerged in the continent. Sorghum and millet were cultivated there, and in Ethiopia early farmers domesticated plants that are still little known outside the region, such as teff, eleusine and korarima, all staples of the Ethiopian diet. In West Africa and the Sahel 4,000 years ago, humans were already changing the landscape. The people who lived in these regions used fire and tree-felling to eliminate undesirable plants and increase land under cultivation, favoured particular trees (shea and locust bean), and hunted both herbivores (antelope and giraffe) and carnivores (lion and hyena).

Our relationship with the grasslands is by no means a thing of the past. At least a third of the world's land surface is made up of grasslands or arid prairies on which over one billion people are dependent and which are, as we saw in the previous chapter, vulnerable to desertification. People are often perceived as a source of problems for natural ecosystems, and it is true that the process of desertification and changes in biodiversity may sometimes be traced to human activities. One culprit is climate change, caused partly by an increase in concentrations of carbon dioxide in the atmosphere and the over-exploitation of agricultural lands, as in the Dust Bowl in 1930s North America. The arrival of the first humans in the Australian bush, some 55,000 years ago, coincided with the extinction of numerous marsupials, including giant species such as Diprotodon, that had lived there in isolation for 50 million years: whether that was cause and effect is a matter of debate.

In the nineteenth and twentieth centuries, ecologists often reacted to this state of affairs by suggesting that pockets of nature should be protected by the exclusion of humans and human activities. By creating national parks and setting up institutions (including paramilitary park rangers) to manage them, governments and organisations dedicated to the protection of nature eliminated the element that was perceived as the problem: humans. The creation of the Hwange National Park in Zimbabwe in the 1920s involved the displacement of the indigenous human populations, who were pushed out to the less fertile lands that surrounded it. Some researchers have calculated that in Africa alone fourteen million people have been displaced in order to create national parks and reserves.

It has become generally accepted that an unbridgeable divide exists between 'nature' and humans, with the latter being condemned as incompatible with the survival of the former. This viewpoint, which has driven the politics of conservation throughout the twentieth century, is now being challenged by an alternative view, which proposes that humans are capable of living in habitats rich in biodiversity without destroying them. First came recognition of the existence of 'traditional ecological knowledge' in numerous human cultures, highly developed systems of understanding that are often possessed by societies that inhabit environments rich in biodiversity. Then a series of historical and archaeological studies showed that many ecosystems once believed to be fundamentally natural were in fact the result of human interaction going back over millennia; these included not only the grasslands of West Africa but also areas of the Amazon rainforest and African equatorial rainforest, which until then had been thought to be virgin. The vogue for creating nature reserves sealed off and hostile to all human activity appears to have lost its impetus, perhaps because by definition they're not tourist-friendly. The current emphasis is on involving local populations in the protection of threatened species, an approach to conservation that, if it works, is likely to be more effective in the long term than measures imposed from outside.

Increasingly attempts are being made to reconcile human food production ecosystems with rich biodiversity. These include agroforestry, in which crops are grown under tree cover and holistic management as seen, for example, in the production methods developed by the Zimbabwean economist Allan Savory. Their proponents claim that these techniques, which include the deployment of livestock, may make it possible to turn deserts green again, countering desertification and supplying a source of food for local people. The resolution of the conflict between nature and farming could be win–win for all concerned.

The grassland ecosystems are particularly pertinent for a consideration of the complex interactions between ourselves and other species. We have lived in these environments for a very long time, and their current desertification poses a major ecological challenge. Our most appropriate response must surely be to accept our responsibilities and to protect the grasslands just as we protect our own dwellings so that both can coexist, as in fact they have done for a long way back into history.

Epilogue: The Zebras and Me

A BATTERED LAND CRUISER leaves the track and edges slowly towards a group of zebras, crushing the dry grass as it creeps forward. The zebras – one male and four females – eye it beadily, ears pricked, vigilant. Then one of them decides the vehicle is too close and moves away. The others follow in its wake.

The Land Cruiser repeats this manoeuvre several times a day, each time with a different group of zebras. The sequence of events barely changes: creep up on the group until one of them becomes too uncomfortable to stay and they move off en masse. Occasionally a subtle variation relieves the monotony: a male comes to inspect the vehicle-threat, or the group bolts off at a gallop, or none of them budges an inch and the vehicle has to stop before it nudges into them.

For a few months I was the driver of that beat-up old Land Cruiser, crisscrossing the dusty tracks of the Hwange National

Park in Zimbabwe. My job was to observe the zebras of the plains in their natural habitat, and from time to time to disturb them a little. The aim was to find out whether plains zebras had individual personalities, to see if there were repeated differences in behaviour between individuals of the species. My task, in other words, was to discover if zebra A was more stressed than zebra B, or if zebra Y was more assertive within the group than zebra Z, and to do this in as many different situations as possible.

Personality in animals emerged as a subject of research a decade or so ago after the discovery that many animals, ranging from species of spider to elephant seals, do indeed possess individual personalities. The area had seemed at first to be relatively trivial but was soon found to have significant consequences for the evolution and ecology of animals: within a few years differences in behaviour provided the key to understanding numerous phenomena, for example:

- How social networks are structured in some species. (According to the personalities of group members.)
- How new environments are colonised. (By the boldest individuals.)
- How sexual partners are chosen (Partly according to the personalities of the male and female.)

In November 2011, when I was looking for a research placement for my masters degree, I was unaware of this scientific research. I was knocking on the door of my future research supervisor, Simon Chamaillé, because I had heard that he conducted frequent field studies abroad and that was what I really wanted to do. To be honest, if travel to some exotic land was part of the deal, I was prepared to work on anything. Simon offered me some research into the personality of zebras that just might – or, as he put it, 'perhaps-possibly-with-a-following-wind' – lead to an experiment in Zimbabwe. That was enough for me: I'd have signed up in blood. I then spent hour upon hour in a lab somewhere in France, analysing interminable videos of zebras. I did this for six months.

My arrival at that lab was a portent of things to come: after cruising through France for several hours at 100kph, one of the tyres on my van exploded. The return journey was no better: first the cambelt broke, more or less destroying the engine, and when eventually I got home I found I'd got a flat battery. Simon had to come and help bump-start the van. That I had overcome such mechanical misfortunes finally convinced him that I was up to going out in the field. Here other clapped-out vehicles awaited me. Eighteen months later I was driving at zebras in a 4x4 that had seen better days and which was undoubtedly a lot older than I was.

When people ask me what I actually did for my research, I tend to simplify. I sometimes say that I spent my time driving vehicles at zebras, which always produces an enjoyable response. Or I say I was trying to understand their behaviour, which has the advantage of being closer to the popular idea of what naturalists are supposed to do in the wild. The real reason was weirder, and would have required more explanation: the truth was I was trying to discover whether zebras' stripes advertised something about their personalities.

The existence of signals on animals' bodies to communicate the presence of a behavioural trait is well known. It was discovered when evolutionary biology was still in its infancy. In insects like wasps, bright colours are used. In mammals, however, the signal is often composed of black and white patches. Honey badgers, for instance, warn of their exceptional levels of aggressiveness through highly conspicuous black-and-white markings. These warning colours are found in many species in temperate zones, such as badgers and skunks. This type of adaptation, designed to deter predators, was discovered by Darwin's contemporary Alfred Russel Wallace and termed 'aposematism' (from the Greek *apo*, away, and *sema*, sign) by the Oxford evolutionary scientist Edward Bagnall Poulton in 1890. It belongs to a very large group of indicators that offer at-a-glance information about the behaviour of the species displaying them.

In the late twentieth century biology moved from being about species to being about individuals. Scientists became more aware of the huge variation in colouring of coats, skin, plumage and the like among a single population. They then observed that these variations were sometimes linked with variations in behaviour: within a given species, individuals with a certain type of personality would tend to display a certain type of colouring.

The genetic origins of this correlation are, in fact, very simple: in some species the gene responsible for production of melanin, the protein responsible for coloration in most mammals, also governs other functions, including stress management, the production of steroid hormones, inflammatory reactions and energy expenditure. When the gene is expressed, it results in a range of behavioural and morphological traits that are all tied together like a bundle. This is a pleiotropic gene, meaning that it has many different effects on the individual that carries it and that, when selection for any one of these effects occurs, the others follow.

The discovery of the genetic basis for this phenomenon prompted numerous studies. From these we have learned, for example, that the size of the black bib on the plumage of the Eurasian siskin reliably indicates an individual's dominance and boldness in exploring its environment. We have also discovered that common kestrels with a larger black tail band are more aggressive than those with smaller, paler tail bands.

We saw in chapter 12 that those lions in the savannah with the darkest manes have the highest levels of testosterone and are the most aggressive. But the phenomenon remains largely unexplored: Simon and I were curious to know whether zebras had personalities and, if they did, whether these could be related in some way to their stripes. The evolutionary origins of zebras' stripes remain shrouded in mystery, but this was not the question we were setting out to address. Starting from the observation that melanin tends to be correlated with behaviour in certain species, we simply wanted to find out whether this was the case with zebras. It was not the pattern of the stripes that interested us but the overall amount of black on the animal.

As I pointed the Land Cruiser doggedly at yet another group of zebras in order to measure their tolerance of disturbance – a common indicator of levels of boldness in animals – the question I was trying to answer was: 'Are the individuals with the finest stripes also the most timid?'

My days unfolded at the uneventful pace of the average zebra day, settling into a well-worn routine: in the late morning I would jump into the Land Cruiser, drive to the park and greet the friendly warden with a nod, head for one of the spots where

I was certain to see zebras, fail to find any, head off for a different spot, find some, get out the video camera, spend ages filming them, eat and drink while I read French comic strips on my laptop, do the driving-towards-them experiment, and repeat.

After a few months, this experience had changed me. I had developed an intense love–hate relationship with the Land Cruiser, I had exhausted my comic collection and I knew everything there was to know about the daily activities of a hundred or so zebras, each of which I now knew individually from the pattern of its stripes. I named them after famous biologists (*'Bonjour*, Hamilton! *Salut*, Hubbell!') or other figures who appealed to me ('Hello, Massoud, still with Battuta?'). I had seen pregnant females with hugely swollen bellies suddenly appear with their newborn foals. I had seen newborn foals disappear. I had seen zebras with hideous wounds, and then watched them heal so well that they were left with only a tiny scar among their stripes. I had been fortunate enough to witness some astonishing scenes of animals in the wild. I was once unlucky enough to have to spend the night out in the bush, just a few tens of metres from a group of hunting lions, but I had been lucky enough to have a clapped-out vehicle to sleep in.

In the end, statistical analysis told me there is not much variation in zebra behaviour, which is to say they do not have much personality. And there were no behavioural signals concealed in their stripes, either. Possibly this is because the environment they live in is too dangerous to admit any variation, however slight, in the finely optimised behaviour that is essential for their survival. Possibly it is because their tightly woven social structure prevents them from developing individual differences.

Or possibly it is because you would need to spend months and months longer observing them in the wild than I did in order to detect any differences or signals.

Whatever the case, it was an unforgettable experience. I encourage any readers inspired by the savannah or by the subjects discussed in this book to go and do something similar. Before you go, though, take a course in car maintenance.

Further Investigations

Chapter 1: The Female Hyena's Penis

A timeline of the development of the human embryo: http://php.med.unsw.edu.au/embryology/index.php?title=Timeline_human_development.

A scientific article on the complexity of the reproductive act in hyenas: Szykman, M. and others, 'Courtship and mating in free-living spotted hyenas', *Behaviour*, 144 (7), 2007, pp. 815–46.

An article offering a survey of the various theories on the function of the pseudo-penis in female hyenas: Muller, M. and Wrangham, R., 'Sexual mimicry in hyenas', *Quarterly Review of Biology*, 77 (1), 2002, pp. 3–16.

A scientific article on the evolution of horns in female bovids: Stankowich, T. and Caro, T., 'Evolution of weaponry in female bovids', *Proceedings of the Royal Society B: Biological Sciences*, 2009, p. 1256, doi: 10.1098/rspb.

Chapter 2: The Giraffe's Long Neck

The scientific article that set the cat among the pigeons by suggesting sexual selection as the principal force responsible for the length of giraffe's necks: Simmons, R. and Scheepers, L., 'Winning by a neck: sexual selection in the evolution of giraffe', *American Naturalist*, 148 (5), 1996, pp. 771–86.

On mortality through necking in giraffes in the Niger: Suraud, J.-P., 'Identifier les contraintes pour la conservation des dernières girafes de l'Afrique de l'Ouest: Déterminants de la dynamique de la population et patron d'occupation spatiale', thesis, Université Claude Bernard-Lyon I, 2011.

But 'there is no evidence that sexual selection was a factor in the evolution of giraffe morphology [or that] the long neck ... evolved as a weapon in males': Mitchell, Graham and others, 'Growth patterns and masses of the heads and necks of male and female giraffes', *Journal of Zoology*, 290 (1), 2013, pp. 49–57.

The scientific article that found universal agreement: Cameron, E. Z. and du Toit, J. T., 'Winning by a neck: tall giraffes avoid competing with shorter browsers', *American Naturalist*, 169 (1), 2007, pp. 130–35.

For a historical perspective on the debate that caused a rift between biologists specialising in giraffes: http://natureinstitute.org/pub/ic/ic10/giraffe.htm.

Above all, this article offering an excellent summary of the debate and the various hypotheses: Wilkinson, David M. and Ruxton, Graeme D., 'Understanding selection for long necks in different taxa', *Biological Reviews*, 87 (3), 2012, pp. 616–30.

And this piece from a French blog: http://ssaft.com/Blog/dotclear/index.php?post/2012/03/07/Sexe%2C-cous-et-Sauropodes.

Chapter 3: The Random Flight of the Gazelle

Alain Pavé's *La Course de la gazelle, biologie et écologie à l'épreuve du hasard*, Editions EDP Sciences, 2011, explores some of the ecological, behavioural and molecular questions referred to here.

A classic work on Protean behaviour was published in 1988: Driver, P. M. and Humphries, D. A., *Protean Behaviour: The Biology of Unpredictability*, Clarendon Press, 1988.

The vinegar fly eye is only one example among many, and sensory systems display many random processes, as discussed (in French) by Claude Desplan in 'Les sens au gré du hasard', *Pour la Science*, 385, 2009, pp. 96–101.

For another interesting and more scientific summary: Losick, R. and Desplan, C., 'Stochasticity and cell fate', *Science*, 320 (5872), 2008, pp. 65–8.

For a summary of the sources of chance in evolution: Malaterre, C. and Merlin, F., 'La part d'aléatoire dans l'évolution. Hasard et incertitudes', *Pour la Science*, 385, 2009, pp. 68–74.

For an explanation (in French) of the importance of the observer effect in particle physics: http://www.youtube.com/watch?v=Cow-gGcrbLE.

On the links between quantum physics and biology: Abbott, D. (ed.), *Quantum Aspects of Life*, World Scientific Publishing, 2008.

And another example of the importance of chance in the living world:

Random effects in cells

Biology is a science that incorporates an understanding of random effects in evolutionary processes. A good example can be found in episodes of mass extinction, random events par excellence, which are clearly destructive but also generate diversity. It seems to me, however, that the 'conceptual revolution in chance' is most applicable at the scale of the invisible, that is at the molecular level, within the cell.

Stochastic (meaning random) processes are also at work inside our cells. To produce proteins, cells transcribe and translate the genes, so that the reading of gene X leads to the production of protein X. For many years, the cell has been likened to a well-oiled machine, with a 'book' containing all the information it needs (deoxyribonucleic acid, or DNA) and assiduous readers (the proteins that read the DNA) who write memos (ribonucleic acid, or RNA) for the benefit of assembly line workers (ribosomes) who read the memos and produce the required proteins. This linear vision of gene expression has sometimes been described as genetic determinism. Each gene produces a specific protein, and the role of random effects is merely to act as a disruptive factor in this rigorous machine, the background noise that interferes with the signal.

Recent discoveries in the field of gene expression contradict this model of cell function: far from being a finely regulated machine, the cell is a constant ferment of molecules subject to thermodynamic agitation which means that molecules show Brownian motion. These random interactions are subjected to processes of selection that introduce the 'order' we can observe. For further information, Jean-Jacques Kupiec has written several books explaining (in French) his views on the subject. The transcript of an interview with him can be read here: http://www.agoravox.fr/actualites/technologies/article/le-chercheur-jean-jacques-kupiec-48970

Chapter 4: How the Zebra Got its Stripes

A summary of the various hypotheses can found in Ruxton, G. D., 'The possible fitness benefits of striped coat coloration for zebra', *Mammal Review*, 32, 2002, pp. 237–44.

On stripes as an anti-fly adaptation: Egri, A. and others, 'Polarotactic tabanids find striped patterns with brightness and/or polarization modulation least attractive: an advantage of zebra stripes', *Journal of Experimental Biology*, 215, 2012, pp. 736–45.

The earliest article to discuss tsetse flies as a selection pressure in the development of stripes: Harris, R. H. T. P., 'Report on the bionomics of the tsetse fly', Provincial Administration of Natal, Pietermaritzburg, South Africa, 1930.

A very recent article using geographical analysis to confirm the fly hypothesis: Caro, T., Izzo, A., Reiner Jr, R. C., Walker, H. and Stankowich, T., 'The function of zebra stripes', *Nature Communications*, 5, 2014, doi:10.1038/ncomms4535.

A recent article on the optical illusions produced by stripes: How, M. J., and Zanker, J. M., 'Motion camouflage induced by zebra stripes', *Zoology*, 117 (3), 2014, pp. 163–70.

And a proposal to paint military vehicles with striped dazzle camouflage against rockets: Stevens, M., Searle, W. T. L., Seymour, J. E., Marshall, K. L. and Ruxton, G. D., 'Motion dazzle and camouflage as distinct anti-predator defenses', *BMC Biology*, 9 (1), 2011, p. 81.

The earliest reference to the idea that zebra stripes might create air fluctuations that cool the animal can be found in Morris, D., *Animal Watching: A Field Guide to Animal Behaviour*, Jonathan Cape, 1990.

The 2015 study that brings all these vexed questions up to date: Larison, B., Harrigan, R. J., Thomassen, H. A., Rubenstein, D. I., Chan-Golston, A. M., Li, E. and Smith, T. B., 'How the zebra got its stripes: a problem with too many solutions', *Royal Society Open Science*, 2 (1), 2015, http://rsos.royalsocietypublishing.org/content/2/1/140452.

Chapter 5: The Air-Conditioning of the Termite Mound

To learn more, go to Scott Turner's excellent and highly informative website: http://www.esf.edu/efb/turner/termitePages/termiteMain.html.

A scientific article on the ecological impact of termite mounds: Pringle, R. M., Doak, D. F., Brody, A. K., Jocqué, R. and Palmer, T. M., 'Spatial pattern enhances ecosystem functioning in an African savanna', *PLoS Biology*, 8 (5), 2010, doi:10.1371/journal.pbio.1000377.

Visit the interior of a termite mound here: http://www.mesomorph.org/.

Chapter 6: The Impala's Mexican Waves

D. J. T. Stumper, the leading authority on collective behaviour, offers an indispensable overview of the subject in 'The principles of collective animal behaviour', *Philosophical Transactions of the Royal Society B: Biological Sciences*, 361 (1465), 2006, pp. 5–22.

Other scientific articles on auto-organisation in collective behaviour:

– Couzin, I. D., Krause J., James, R., Ruxton, G. D. and Franks, N. R., 'Collective memory and spatial sorting in animal groups', *Journal of Theoretical Biology*, 218, 2002, pp. 1–11.

– Helbing, D., Farkas, I. and Vicsek, T., 'Simulating dynamical features of escape panic', *Nature*, 407 (6803), 2000, pp. 487–90.

– Kuramoto Y., *Chemical Oscillations: Waves and Turbulence*, Springer, 1984.

– Néda, Z., Ravasz, E., Brechet, Y., Vicsek, T. and Barabási, A. L., 'The sound of many hands clapping', *Nature*, 403 (6772), 2000, pp. 849–50.

– Pays, O., and others, 'Prey synchronize their vigilant behaviour with other group members', *Proceedings of*

the Royal Society B: Biological Sciences, 274 (1615), 2007,
pp. 1287–91.

Chapter 7: Elephant Dictatorship vs Buffalo Democracy

A good summary of group decision-making among animals
can be found here: Conradt, L. and Timothy, J. R., 'Consensus
decision-making in animals', *Trends in Ecology & Evolution*, 20
(8), 2005, pp. 449–56.

I also liked this little study by a researcher in political science:
List, C., 'Democracy in animal groups: a political science
perspective', *Trends in Ecology & Evolution*, 19 (4), 2004, pp. 166–
8.

A talk on the wisdom of crowds, recorded at the Collège de
France: http://www.canalu.tv/video/college_de_france/
microfoundations_of_collective_wisdom.4046.

And a book by Scott E. Page: *The Difference: How the Power
of Diversity Creates Better Groups, Firms, Schools, and Societies*,
Princeton University Press, 2007.

On buffalo voting: Prins, H. H. T., *Ecology and Behaviour of the
African Buffalo*, Chapman & Hall, 1996.

On elephant matriarchs: McComb, K., Shannon, G.,
Durant, S. M., Sayialel, K., Slotow, R., Poole, J. and Moss,
C., 'Leadership in elephants: the adaptive value of age',
Proceedings of the Royal Society B: Biological Sciences, 278 (1722),
2011, pp. 3270–76.

On a very simple model explaining the emergence of a
leader when there is a difference in needs on one hand and
a desire to preserve group cohesion on the other: Rands, S.,

Cowlishaw, G. and Pettifor, R., 'Spontaneous emergence of leaders and followers in foraging pairs', *Nature*, 423 (6938), 2003, pp. 432–4.

For an understanding of why the accuracy of a crowd increases with its diversity, an illustration of the theory and diversity of predictions:

Diversity and Prediction

The diversity prediction theorem is a simple matter of statistics. For seasoned statisticians, this is 'merely' a direct reformulation of what is called the bias-variance trade-off. For normal people, the best thing is to 'feel' how this works, and for this we need to go to the market.

Imagine that a couple called Julia and Tony are arguing over their estimates of the weight of three different vegetables. So let us have a look at how their combined judgement is more reliable than their individual guesses.

The actual weight of the carrots is 6kg, the aubergines weigh 5kg and the potatoes 1kg. These are the absolute values that Julia and Tony are trying to estimate.

For the carrots, Julia estimates 6kg and Tony 10kg; for the aubergines, they guess 3kg and 7kg, respectively; and for the potatoes, they guess 5kg and 1kg.

It is worth noting that for the potatoes Julia is 4kg wide of the mark. If we calculate the average of Julia and Tony's guesses, we arrive at their collective estimate, thus:

Carrots: (6 and 10), giving an average of 8
Aubergines: (3 and 7), giving an average of 5
Potatoes: (5 and 1), giving an average of 3

It is these averages that are supposed to be more

---

accurate than the couple's individual estimates. We can check this by working out (1) the average margin of error of the two participants, and (2) the margin of error of their collective estimate.

(1) The participants' margin of error
To work this out we need to start by calculating the margin of error of each participant, that is, the difference between their estimate and the actual value. Then we square it to give equal weight to the positive and negative differences. Julia was okg out for the carrots, 2kg out for the aubergines, and 4kg out for the potatoes. So the sum of her errors (squared) is as follows:

$$(6-6)^2 + (3-5)^2 + (5-1)^2 =$$
$$(0)^2 + (-2)^2 + (4)^2 = 0 + 4 + 16 = 20$$

The same calculation for Tony gives:

$$(10-6)^2 + (7-5)^2 + (1-1)^2 = 16 + 4 + 0 = 20$$

(2) Their collective error
We can use the same principle to work out their collective error:

$$(8-6)^2 + (5-5)^2 + (3-1)^2 = 4 + 0 + 4 = 8$$

So the margin of error of their collective estimate is 8. Julia and Tony's average margin of error is 20. So there is a difference of 12 between the two, and their collective error is smaller. But where does this 12 come from? From the diversity prediction theorem. As Scott E. Page explains, the collective error (8) equals Julia and Tony's average error (20) minus the diversity prediction (12).

This diversity prediction is very easy to find: it is the difference between the estimates of the participants and the overall average of their estimates (also known as the variance). For the carrots, the average of the estimates was higher, at 8. For the aubergines it was 5. For the potatoes, 3. Julie's divergence from this average is:

$$(6-8)^2 + (3-5)^2 + (5-3)^2 = 4 + 4 + 4 = 12$$

Tony's is:

$$(10-8)^2 + (7-5)^2 + (1-3)^2 = 4 + 4 + 4 = 12$$

Their average is therefore the diversity prediction: the average of (12 and 12) is 12.

Page goes on to explain that as there is always some diversity, the collective error will always be smaller than the average error of each participant (from which the diversity is therefore subtracted). In other words, as soon as there are people with different opinions, their collective accuracy will be greater than the average accuracy of any individual. QED. Congratulations for sticking with it to the end.

Chapter 8: The Antelope Art of Sexual Manipulation

The original article on manipulative topi antelopes: Bro-Jørgensen, J. and Pangle, W. M., 'Male topi antelopes alarm snort deceptively to retain females for mating', *American Naturalist*, 176 (1), 2010, E33–9.

Chapter 9: Dung Beetle Navigation

The article on the experiment with dung beetles and the Milky Way: Dacke, M. and others, 'Dung beetles use the Milky Way for orientation', *Current Biology*, 23 (4), 2013, pp. 298–300.

The article studying the way humans walk when lost in environments without distinct visual clues: Souman, J. L., Frissen, I., Sreenivasa, M. N. and Ernst, M. O., 'Walking straight into circles', *Current Biology*, 19 (18), 2009, pp. 1538–42.

Chapter 10: Seismic Signalling in the Elephants' Sound-World

Karen McComb's work showing elephants can distinguish between Maasai and Kamba people: McComb, K. and others, 'Elephants can determine ethnicity, gender, and age from acoustic cues in human voices', *Proceedings of the National Academy of Sciences*, 111 (14), 2014, pp. 5433–8.

How elephants and their extinct relatives have been massively hunted as *Homo sapiens* has spread over the world's continents: Sanchez, G. and others, 'Human (Clovis)-gomphothere (Cuvieronius sp.) association~ 13,390 calibrated yBP in Sonora, Mexico', *Proceedings of the National Academy of Sciences*, 111 (30), 2014, pp. 10972–7; Surovell, T., Waguespack, N. and Brantingham, P. J., 'Global archaeological evidence for proboscidean overkill', *Proceedings of the National Academy of Sciences*, 102 (17), 2005, pp. 6231–6.

On the highly developed acoustic social network shared by elephants: McComb, K. and others, 'Unusually extensive networks of vocal recognition in African elephants', *Animal Behaviour*, 59 (6), 2000, pp. 1103–9.

Listen to elephants' wide vocal repertoire on the *National Geographic* website http://www.nationalgeographic.com/

news-features/what-elephant-calls-mean/ or on the official site of Elephant Voices http://www.elephantvoices.org/.

Poole, J. H., 'Behavioral contexts of elephant acoustic communication', in Moss, Cynthia J., Croze, Harvey and Lee, Phyllis C., eds., *The Amboseli Elephants: A Long-term Perspective on a Long-lived Mammal*, University of Chicago Press, 2011, pp. 125–61.

Caitlin O'Connell's articles are the authority on seismic communication among elephants, as her work has laid the foundations in this field: O'Connell-Rodwell, C. E., 'Keeping an "ear" to the ground: seismic communication in elephants', *Physiology* (Bethesda), 22, 2007, pp. 287–94.

Chapter 11: Honey Badger – Weapon of Mass Destruction

This remarkable species has its own dedicated website: http://www.honeybadger.com/.

This chapter was influenced by Ben Thompson, whose brilliant historical biographies can be found on his website, www.badassoftheweek.com.

For some of recent research on badger resistance to snake poison, see: http://www.slate.com/blogs/wild_things/2015/06/16/honey_badger_venom_resistance_biologists_discover_the_secret.html.

For a sceptical perspective on badgers and honeyguides see: http://blogs.discovermagazine.com/notrocketscience/2011/09/19/lies-damned-lies-and-honey-badgers/#.VsOKYhE3vRc.

The honey badger's exploits can be found on the internet, e.g. at: http://www.youtube.com/watch?v=4r7wHMg5Yjg.

On gangs of honey badgers attacking the honest citizens of Basra in Iraq see: http://www.news.com.au/dailytelegraph/story/0,22049,22056684-5001028,00.html.

For a mutant hybrid of grizzly bear, great white shark and giant quid, see the bearsharktopus: http://knowyourmeme.com/memes/bearsharktopus.

The South African Ratel infantry fighting vehicle: http://en.wikipedia.org/wiki/Ratel_IFV.

Chapter 12: The Truth about the Lion King

A study showing that lions with dark manes have higher levels of testosterone than those with light manes: West, P. and Packer, C., 'Sexual selection, temperature, and the lion's mane', Science, 297 (5585), 2002, pp. 1339–43. http://www.sciencemag.org/content/297/5585/1339.short.

Wikipedia, some examples of hyena calls: http://en.wikipedia.org/wiki/Spotted_hyena#Vocalisations.

The consanguinity of the Ngorongoro Crater lions is clearly explained here: http://fish-dont-exist.blogspot.fr/2013/01/les-lions-ce-que-vous-napprendrez-pas.html, and is the subject of this scientific article: Wildt, D. E., Bush, M., Goodrowe, K. L., Packer, C., Pusey, A. E., Brown, J. L., Joslin, P. and O'Brien, S. J., 'Reproductive and genetic consequences of founding isolated lion populations', Nature, 329 (6137), 1987, pp. 328–31.

Chapter 13: How to Turn a Lion into a Cub-Killer

A brief history of the pumped, man-made waterholes in Hwange National Park in Zimbabwe: http://newswatch. nationalgeographic.com/2013/03/08/waterholes-hwange-national-park-zimbabwe/.

Vintage BBC footage of blue tit and great tit 'milk thieves': http://www.youtube.com/watch?v=Svozh7a1_p4.

A scientific article on black kites and plastic nests: Fabrizio, S. and others, 'Raptor nest decorations are a reliable threat against conspecifics', *Science*, 331 (6015), 2011, pp. 327–30.

A scientific study of the use of cigarette butts by birds: Suárez-Rodríguez, M., López-Rull, I. and Macias García, C., 'Incorporation of cigarette butts into nests reduces nest ectoparasite load in urban birds: new ingredients for an old recipe?', *Biology Letters*, 9 (1), 2012, doi:10.1098/rsbl.2012.0931.

The BBC and David Attenborough demonstrate how crows use cars to crack nuts: https://www.youtube.com/watch?v=BGPGknpq3eo.

And a scientific article studying the spread of this phenomenon: Yoshiaki, N. and Higuchi, H., 'When and where did crows learn to use automobiles as nutcrackers?', *Tohoku Psychologica Folia*, 60, 2002, pp. 93–7.

On birds and speed limits: Legagneux, P. and Ducatez, S., 'European birds adjust their flight initiation distance to road speed limits', *Biology Letters*, 9 (5), 2013, doi:10.1098/rsbl.2013.0417.

DeVault, T. L., Blackwell, B. F., Seamans, T. W., Lima S. L. and Fernández-Juricic, E., 'Speed kills: ineffective avian escape responses to oncoming vehicles', *Proceedings of the Royal Society B*, 282, 2015, http://rspb.royalsocietypublishing.org/content/282/1801/20142188.

A pigeon surfing on a car: http://www.youtube.com/watch?v=8BooGKDv-Zg.

The scientific article behind the frogs singing in drains: Tan, W. H. and others, 'Urban canyon effect: storm drains enhance call characteristics of the Mientien tree frog', *Journal of Zoology*, 294 (2), 2014, pp. 77–84.

For the impact of trophy hunting on levels of infanticide, see this research paper: Loveridge, A. J. and others, 'The impact of sport-hunting on the population dynamics of an African lion population in a protected area', *Biological Conservation*, 134 (4), 2007, pp. 548–58.

For more on the life of Cecil the great Hwange lion see: http://friendsofhwange.com/twisting-tales-cecil-and-jericho/ and on his death see: https://en.wikipedia.org/wiki/Killing_of_Cecil_the_lion.

On the social effects of elephant culling: Graeme, S. and others, 'Effects of social disruption in elephants persist decades after culling', *Frontiers in Zoology*, 10 (1), 2013, p. 62.

For a more detailed account of changing points of view on the relationship between humans and their environment: http://thebreakthrough.org/index.php/programs/conservation-and-development/humanitys-pervasive-environmental-influence-began-long-ago/.

Chapter 14: Catastrophic Change

A fuller exploration of the mechanics of catastrophic transitions can be found in an excellent and very clear discussion (in French) by Sonia Kéfi on the equally excellent *Regards sur la biodiversité* site of the Société française d'écologie: http://www.sfecologie.org/regards/2012/10/19/r37-hysteresis-sonia-kefi/.

More examples (in French) of catastrophic transitions in ecosystems: http://danslestesticulesdedarwin.blogspot.it/2013/05/avalanches-sur-la-biosphere-episode-2.html.

More detailed examinations in other scholarly articles:

- Rietkerk, M., Dekker, S. C., De Ruiter, P. C. and Van de Koppel, J., 'Self-organized patchiness and catastrophic shifts in ecosystems', *Science*, 305 (5692), 2004, pp. 1926–9.

- Scheffer, M., Bascompte, J., Brock, W. A., Brovkin, V., Carpenter, S. R., Dakos, V., Held, H. and others, 'Early warning signals for critical transitions', *Nature*, 461 (7260), 2009, pp. 53–9.

- Scheffer, M., Carpenter, S., Foley, J. A., Folke, C. and Walker, B., 'Catastrophic shifts in ecosystems', *Nature*, 413 (6856), 2001, pp. 591–6.

- Scheffer, M., Carpenter, S. R., Lenton, T. M., Bascompte, J., Brock, W., Dakos, V., Van de Koppel, J. and others, 'Anticipating critical transitions', *Science*, 338 (6105), 2012, pp. 344–8.

Chapter 15: Human Evolution and its Impact

Wikipedia has a very clear and well-referenced chronology of human evolutionary history, highlighting the significant events in biological evolution that would one day produce *Homo sapiens*: http://en.wikipedia.org/wiki/Timeline_of_human_evolution.

The use of cooking by *Homo erectus* appears to date back at least two million years; the results are published in this article: Organ, C., Nunn, C. L., Machanda, Z. and Wrangham, R. W., 'Phylogenetic rate shifts in feeding time during the evolution of *Homo*', *Proceedings of the National Academy of Sciences*, 108 (35), 2011, pp. 14555–9.

But this has also attracted vociferous criticism: Roebroeks, W. and Villa, P., 'On the earliest evidence for habitual use of fire in Europe', *Proceedings of the National Academy of Sciences*, 108 (13), 2011, pp. 5209–14.

A good survey of the evolution of intelligence and brain size in primates (including us): Gerhard, R. and Dicke, U., 'Evolution of the brain and intelligence', *Trends in Cognitive Sciences*, 9 (5) 2005, pp. 250–57.

Homo erectus followed elephants for many years: Ben-Dor, M., Gopher, A., Hershkovitz, I. and Barkai, R., 'Man the fat hunter: the demise of *Homo erectus* and the emergence of a new hominin lineage in the Middle Pleistocene (ca. 400 kyr) Levant', *PLoS One*, 6 (12), 2011, doi:10.1371/journal.pone.0028689.

Homo sapiens is over 300,000 years old, according to recent genetic results: Mendez, F. L., Krahn, T., Schrack, B., Krahn,

A. M., Veeramah, K. R., Woerner, A. E. and Hammer, M. F., 'An African American paternal lineage adds an extremely ancient root to the human Y chromosome phylogenetic tree', *American Journal of Human Genetics*, 92 (3), 2013, pp. 454–9.

A very recent article on genes under strong selection that have come down to us from Neanderthal man: Sankararaman, S., Swapan, M., Dannemann, M. and others, 'The genomic landscape of Neanderthal ancestry in present-day humans', *Nature*, 507 (7492) 2014, pp. 354–7.

A popular article (in French) on the grasslands of West Africa as ecosystems influenced by humans for millennia: Ballouche, A. and Rasse, M., 'L'homme, artisan des paysages de savane', *Pour la science*, 358, 2007, pp. 56–61.

On conservation refugees: Dowie, M., *Conservation Refugees: The Hundred-Year Conflict between Global Conservation and Native Peoples*, MIT Press, 2009, p. 12.

An unusually interesting article on the ideology of conservation in the twentieth century: Kareiva, P., Marvier, M. and Lalasz, R., 'Conservation in the Anthropocene: beyond solitude and fragility', *The Breakthrough Journal*: http://thebreakthrough.org/index.php/journal/past-issues/issue-2/conservation-in-the-anthropocene.

For more on the views of Allan Savory, watch his TED talk: www.ted.com/talks/allan_savory_how_to_green_the_world_s_deserts_and_reverse_climate_change.html.

Acknowledgements

This book would not have happened without the generous support of many people. Thank you to my friends Alice Baniel, Timothée Bonnet, Paul Saunders and Pascal Milesi; to Caroline Grou, the Canet de Kerdour family, Jean-Yves Delaveux and Ofelia Cruces; to the Gremion family, and above all to my family, Éric Vitale, Jean-Marie Dadolle, Julia and my parents.

Among the inspirations present throughout the text are three outstanding blogs: elements of Pierre Kerner's *Strange Stuff and Funky Things*, Xochipilli's *Le Webinet des curiosités*, and Tom Roud's *Matières vivantes* can be found in chapters 2, 6 and 3, respectively. All their blogs are mines of information, and they are some of those who have motivated me in popularising science.

Thank you to Colas for agreeing to do the illustrations, and to Gwen, for all the rest.

Index